T0344526

THE COSMIC MICROWAVE BACKGROUND

This volume tells the untold story of how observations of the cosmic microwave background radiation were interpreted in the decades following its serendipitous discovery before the Hot Big Bang model became the accepted orthodoxy. The authors guide the reader through this history, including the many false trails and blind alleys that occurred along the way. Readers will discover how the Big Bang theory was shaped by alternative theories that exposed its weaknesses – including some that persist even today. By looking carefully at what it takes to reject an incorrect theory and the assumptions and processes at each stage, the authors examine the epistemological factors at play between an emerging scientific orthodoxy and alternatives. Their analysis of the cosmic microwave background provides a uniquely well-documented case study of theory building for a wide readership spanning cosmology, the history of physics and astronomy, and the philosophy of science more broadly.

SLOBODAN PEROVIĆ is a professor of philosophy and history of science at the University of Belgrade. He earned his Ph.D. at York University, Toronto, in 2005, and held teaching and research positions at Carleton University and the University of Pittsburgh, before returning to join the faculty in Belgrade in 2010. He is the principal convener of the "Philosophy of Scientific Experimentation" conference series, and is the author of *From Data to Quanta: Niels Bohr's Vision of Physics* (2021).

MILAN M. ĆIRKOVIĆ is a senior researcher at the Astronomical Observatory of Belgrade. He also holds an adjunct position at the Future of Humanity Institute, University of Oxford. He has broad research interests in applying philosophical ideas to diverse areas including astrobiology and global catastrophic risks. His previous books include *The Astrobiological Landscape: Philosophical Foundations of the Study of Cosmic Life* (Cambridge University Press, 2012) and *The Great Silence: Science and Philosophy of Fermi's Paradox* (2018).

THE COSMIC MICROWAVE BACKGROUND

Historical and Philosophical Lessons

SLOBODAN PEROVIĆ

University of Belgrade

MILAN M. ĆIRKOVIĆ

Astronomical Observatory of Belgrade

CAMBRIDGE
UNIVERSITY PRESS

Shaftesbury Road, Cambridge CB2 8EA, United Kingdom

One Liberty Plaza, 20th Floor, New York, NY 10006, USA

477 Williamstown Road, Port Melbourne, VIC 3207, Australia

314–321, 3rd Floor, Plot 3, Splendor Forum, Jasola District Centre, New Delhi – 110025, India

103 Penang Road, #05–06/07, Visioncrest Commercial, Singapore 238467

Cambridge University Press is part of Cambridge University Press & Assessment, a department of the University of Cambridge.

We share the University's mission to contribute to society through the pursuit of education, learning and research at the highest international levels of excellence.

www.cambridge.org
Information on this title: www.cambridge.org/9781108844604

DOI: 10.1017/9781108951968

First published 2024

A catalogue record for this publication is available from the British Library

Library of Congress Cataloging-in-Publication Data
Names: Perović, Slobodan (Philosopher and historian of science), author. | Ćirković, Milan M., author.
Title: The cosmic microwave background : historical and philosophical lessons / Slobodan Perović, University of Belgrade, Milan M. Ćirković, Astronomical Observatory of Belgrade, Serbia.
Description: Cambridge, United Kingdom ; New York, NY : Cambridge University Press, 2024. | Includes bibliographical references and index.
Identifiers: LCCN 2023057747 | ISBN 9781108844604 (hardback) | ISBN 9781108951968 (ebook)
Subjects: LCSH: Cosmic background radiation. | Cosmology – Philosophy.
Classification: LCC QB991.C64 .P47 2024 | DDC 523.1–dc23/eng/20240109
LC record available at https://lccn.loc.gov/2023057747

ISBN 978-1-108-84460-4 Hardback

What is opposed unites, and the most beautiful harmony arises from things that carry in opposite directions; all things come to be by strife.

(Heraclitus of Ephesus)[1]

The astronomer's problem is not a lack of information but an embarrassing excess of it.

(Fred Hoyle)[2]

Contents

Acknowledgments

We would like to express our gratitude to the audiences at the Super-PAC workshop on the philosophy of astrophysics and cosmology, which was held at the Center for Philosophy of Science (University of Pittsburgh) in 2017, as well as to the Institute for Philosophy at the Hungarian Academy of Sciences, the Ian Ramsey Centre (Oxford) and the Future of Humanity Institute (Oxford). Special acknowledgments to Allan Franklin and Dejan Urošević for their feedback on the final draft, John Norton and Helge Kragh for their valuable early suggestions, Vesna Milošević-Zdjelar, Branislav Nikolić, Anders Sandberg, Aleksandar Obradović, and Dušan Pavlović for their assistance in obtaining references, as well as late Petar Grujić, late Jelena Milogradov-Turin, Miroljub Dugić, Srđan Samurović, Richard Dawid, Sir Martin Rees, Milan Stojanović, Marko Stalevski, late Amos Zahiland and numerous other teachers and colleagues for engaging discussions. We extend our thanks to the CUP editors for their patience and diligence, and Elizabeth Thompson and Nada Dumić for their editorial support. This research was supported by the Templeton Foundation and the Ministry of Education, Science and Technological Development of the Republic of Serbia. Special appreciation to Milica Ćirković, Ljiljana Radenović, Klara Perović – and finally, Mowgli the cat and Kruško the dog.

Introduction

Our study of the explanation of the cosmic microwave background (CMB) radiation – the uniform microwave radiation omnipresent across the skies – is philosophically motivated historical analysis. The analysis tells us what viable alternatives to the explanation of the CMB based on the Hot Big Bang model that eventually became the standard CMB interpretation were discussed or neglected after Arno Penzias and Robert Wilson's discovery of the CMB and clarifies their exact role in the developing consensus. The usual impression of a quick consensus ignores multiple methodologically sound alternative explanatory hypotheses of the CMB, now mostly forgotten, but it disappears as soon as we dive into them. This widespread impression has prevented the completion of an adequate detailed picture of both the history and the methodology of cosmology and also precluded the ability to draw some important historic-philosophical lessons relevant to contemporary cosmological research. Even more importantly, a source of potentially valuable ideas may have been sidelined.

In this book, we seek to weave a historical tapestry of this amazing development by considering some forgotten approaches and those presently deemed peripheral and to draw from it methodological and philosophical lessons for modern cosmology. The general motivation for this study is perhaps best expressed by Helge Kragh's comments on the history of cosmology:

[t]here is the tendency to streamline history and ignore the many false trails and blind alleys that may seem so irrelevant to the road that led to modern knowledge. It goes without saying that such streamlining is bad history and that its main function is to celebrate modern science rather than obtain understanding of how science has really developed. The road to modern cosmology abounded with what can now be seen were false trails and blind alleys, but at the time were considered to be significant contributions. (Kragh, 1997, 67–68)

This comment nicely sums up the situation with the alternative interpretations of the CMB. It is a paradigmatic story in this respect. Many scientists and popularizers of science use every opportunity to hail the orthodox interpretation of the CMB as one of the greatest triumphs, if not the greatest one, of modern cosmological science. This is certainly justified, but it blurs the distinction between the actual physical phenomenon and the role the now dominant interpretation played historically. It is as if the CMB photons themselves ended the great cosmological controversy of the 1950s and 1960s (Chapter 2), and there was no long and arduous process of ever-increasing consensus on the emerging and constantly developing orthodoxy. One way to remedy this situation is to carefully analyze the nonorthodox interpretations offered by prominent physicists and cosmologists at the time – sometimes by those who were developing the emerging orthodox view.

In fact, all these concerns go back to the complex and insufficiently studied problematic of paradigm formation in modern cosmology (Kragh, 1997; Norton, 2017). Our historical case study of the formation of the alternatives in modern cosmology sets the scene for our assessment of their respective epistemic standing, primarily with respect to their interpretation of the CMB (Parts III and VI). Understanding the exact epistemological role the CMB has played in modern cosmology is also essential if we wish to come up with a substantial response to broad criticisms of cosmology as a scientific field (e.g., Dingle, 1954; Disney, 2000).[3] This sort of criticism has been around throughout the twentieth and twenty-first centuries; it is still alive and aimed at current cosmological endeavors. It prompts us to ask, for instance, how we can justifiably draw predictions based on high-precision models of the physical state of the early universe and the observed corresponding traces if we do not fully understand the methodological premises on which the alternatives were refuted in the case of the CMB. The methodological premises and desiderata have hardly changed since then, despite a more precise observational nexus. The following question remains to be answered as well: What exactly supports extrapolating to the states of matter many orders of magnitude more extreme than anything we encounter in a laboratory? Understanding the emergence and acceptance of the standard CMB interpretation and the rejection of the alternatives will help us answer these general questions on the methodological foundations of current cosmology.

First, this analysis may yield a more accurate historical understanding of how scientific process proceeded and its epistemological and methodological ramifications. It turns out that the alternative explanations of the CMB were surprisingly varied, ranging from cosmological explanations set firmly within the theoretical framework of the general theory of relativity, to purportedly nonrelativistic cosmological explanations and non-cosmological ones relying on regular physical laws alone. Moreover, this episode in the history of modern cosmology offers a model of epistemic and methodological responsibility in generating alternative explanations in the context of a gradually emerging orthodox account predicated on the constantly improving yet indecisive observational results (i.e., during a prolonged underdetermination of competing theories and models based on available evidence). As such, this episode offers epistemological-methodological lessons for contemporary cosmology, including the role of broader epistemological and metaphysical views, and suggesting viable institutional structures that can facilitate an adequate playing field for efficient progress.

Second, the alternative theories may be a source of still-useful conjectures and ideas. This may be a worthwhile topic in its own right. With this in mind, we discuss some early ideas about baryonic matter in the cosmological context, assumptions about isotropy in the early universe, and the fields generating the expansion of the universe. Discovering relevant ideas in abandoned or ignored theories in the history of modern physics is not as uncommon as we may think. For example, the central concepts of Machian theories of gravitation, such as Brans–Dicke theory (e.g., Dicke, 1962) and Wheeler–Feynman action-at-a-distance classical electrodynamics (Wheeler & Feynman, 1945, 1949; Hoyle & Narlikar, 1964, 1971; Hogarth, 1962) were seen as too radical at their inception. Yet these concepts are now debated in inflationary cosmology and in philosophical discussions of the arrow of time (Linde, 1990; Price, 1991). In fact, overlooking alternatives may slow down the progress.[4]

This reassessment will inevitably lead us to a set of questions concerning the present situation: How convincing is the standard account currently? Are there any viable alternatives now? If not, why not, and is the critical examination in the modern practical work satisfying? We tackle these questions throughout the book but especially in the closing chapters.

Our discussion starts with a brief introduction to the controversial beginnings of physical cosmology (Part I). We move on to the characterization of the current orthodox interpretation of the CMB predicated on the emerging precision cosmology (Part II) and discuss epistemological and methodological ramifications of the formation of the orthodox view and alternatives (Part III). Finally, we analyze both moderate (Part IV) and radical (Part V) alternatives to the orthodox view and draw potentially far-reaching lessons (Parts VI and VII).

Part I

Physical cosmology: A brief introduction

1

Physical cosmology from Einstein to 1965

Several histories of physical cosmology have been written so far, some of them superb instances of history of science writing (e.g., North, 1994; Kragh, 1996, 2004, 2012; Kragh and Longair, 2019; Peebles, 2020), but to set the stage for what follows, we will recapitulate a few points salient for our study. The treatment will out of necessity be very brief; all interested readers are directed to the appropriate items in the literature. Many good textbooks give the physics side of the story; some contain selected historical analyses and reminiscences. A good example is Peebles (1993).

Modern physical cosmology "officially" started with Albert Einstein's 1917 paper, although a good case could be made for the Boltzmann–Zermelo debate about 20 years earlier, as it was the first debate to explicitly invoke the beginning of the universe within the discourse of physics, and it appeared in major peer-reviewed research journals.[5] The debate focused on the entropy of the universe and Ludwig Boltzmann's anthropic strategy for explaining the thermodynamical arrow of time and, as such, could be regarded as relevant for the cosmological *boundary conditions* (see the rehashing of the anthropic strategy for modern day in Ćirković, 2003). Any hope of understanding the *dynamical laws* had to wait for the development of Einstein's theory of gravitation.

General relativity, however, performed admirably in setting the groundwork for cosmological dynamics in the true sense, even if its first two applications were, somewhat ironically, *static*. In the 100+ years since Einstein's first cosmological model, physical cosmology has radically transformed our view of the world, perhaps the most radical such shift since Nicolaus Copernicus and Galileo Galilei. For present purposes, we can divide its rich history up to 1965 into four major eras.

The first era comprised a sort of prehistory of cosmology, stretching from antiquity to 1917. Among the important early elements, we might mention Olbers's paradox and the so-called gravitational paradox originating with Isaac Newton, both of which seemed to indicate the finite age of the physical world.[6] In the last part of this period, a very interesting cosmological debate was ignited by Ludwig Boltzmann in his controversy with Ernst Zermelo (1895–1896) on the origin of the thermodynamical disequilibrium of the universe – attested to, as Boltzmann presciently argued, by the existence of us as intelligent observers. The tremendous improvements in astronomical instrumentation in the course of the nineteenth century and throughout the first decades of the twentieth century culminated in the construction of the 100-inch (2.5 m) Hooker telescope at Mount Wilson Observatory, California, dedicated in 1917 – the year of the first cosmological models. Before long, the Mount Wilson telescope ushered in a kind of Copernican turn within the realm of physical cosmology.

The second era featured the pursuit of the static universe and lasted from 1917 to 1929. Two main early models, Einstein's static universe and de Sitter's empty universe, emerged in 1917 as a consequence of the great theoretical breakthrough in formulating the first and still the best metric theory of gravity. Both models were characterized by contemporaries as static, based on the mathematical fact that their metric coefficients are independent of time, although the label subsequently only applied to the Einstein model. The dearth of empirical knowledge continued in this phase, and cosmology was regarded as a mathematical game rather than a description of physical reality (and published in corresponding sections of scientific journals). The seminal work of Alexander Friedmann in the early 1920s was not widely known and was rediscovered only later.

The third era developed the expanding mathematical universe (1929–1948). Edwin Hubble and Milton L. Humason's discovery of the expanding universe created conditions for a "cosmological revolution" – cosmologists finally had something clearly dynamical to work with. Georges Lemaître (1931) and George Gamow (1946, 1948, 1949), together with the belatedly read and understood contribution of Alexander Friedmann (1922), laid the foundations for relativistic cosmological models of the expanding universe. What Lemaître called the "primeval atom" became better known as the (hot) Big Bang. The work of those three, together with the important input of Sir Arthur Eddington, Richard Tolman, and a few others who shyly started calling themselves cosmologists,[7] helped physical cosmology attain a modicum of seriousness and authority. Today, this era looks and is often described, especially in popular accounts, as heroic, but we should keep in mind that cosmology was still, in Fred Hoyle's pointed words, the science of "two and a half facts."[8]

The final era, the "great controversy," raged from 1948 to 1965 and merits its own chapter.

2

The "great controversy" (1948–1965) and epistemological issues it raised

With the emergence of the classical steady-state theory in 1948, cosmology entered a new phase of accelerated development in terms of both its theoretical and its observational aspects. This phase was characterized by intense rivalry between the steady-state theory and the family of relativistic world models (which during this time obtained the initially pejorative label of "Big Bang" in a BBC radio program hosted by Hoyle). This rivalry provoked both the crystallization (out of vague speculation) of key tenets of cosmology in serious, well-defined, and at least potentially testable forms, and the creation of crucial observational tests of world models in order to resolve the controversy. A huge amount of philosophical debate took place in those years, as described by Helge Kragh in his magisterial *Cosmology and Controversy* (Kragh, 1996), but this should not detract from the fact that some of the key observational tests of world models were conceived (and some even executed in practice) in this period. This, of course, included quantitative modeling of issues, such as star-formation histories, duration of activity of radio sources, and other evolutionary effects in galaxies; for the first time, these became necessary ingredients instead of *ad hoc* speculations. Here, as elsewhere, the past was prologue. The advent of the classical steady-state theory (we use this term to distinguish it from subsequent "revised" versions and other modifications) was a watershed moment, leading to the emergence of empirical standards fit to discriminate between theories in physical cosmology. More than any other dominant paradigm in any science, the standard cosmological model (Hot Big Bang) has emerged victorious in the context of challenging circumstances, controversies, and conflict. This, we may further speculate, is the reason why its acceptance *outside* academic research has been so slow, uneven, and in many ways, problematic. And this is why we have so many amateurish "revolutionaries" in cosmology compared, say, to chemistry or geophysics or even climate science.

Mesmerized by the latest observational insights into the origin of the universe, disseminated widely through media and social networks on a daily basis, we tend to forget that the treatment of cosmology as a proper scientific subject is a relatively recent occurrence, both in the scientific community and among the general public. Whether cosmological questions were a proper scientific subject was open for debate or even outright dismissed as an inherently unscientific query as late as the 1930s and mid-1940s. The skepticism about the scientific treatment of the origin of the physical universe was a regular textbook attitude.[9] Although it has abated, skepticism is still alive within the scientific community and part of a larger dismissal of historical explanations of natural and social phenomena as ultimately unscientific.[10]

A prominent figure among physicists who dismissed the search for the origins of the universe as inherently unscientific was Herbert Dingle (1953). His dismissive attitude was by no means an exception; any attempts to suggest or model how the universe was or is currently created were guaranteed to meet with severe rebuttals. F. Hoyle noted that the astronomical community in the late 1930s and early 1940s was much more intellectually conservative than that of physics. The latter was happy to speculate and offer exploratory ideas, but the former was not (Gregory, 2005, 16). In fact, Hoyle argued this was a sign of a changing organizational structure and culture of doing science in general; it proved to be an additional obstacle to modern cosmology, leading it along less risky but also less fruitful roads (Gregory, 2005, 337).

In the period leading up to 1965, the so-called (neo)classical cosmological tests were devised: epistemically parsimonious observational criteria for discriminating between rival cosmological models.[11] One was the age of the universe – historically a major motivation for proposing the classical steady-state theory, at least until 1952 and Walter Baade's revision of the cosmological distance scale. Baade brought the predictions of various Big Bang models into at least a vague agreement with the constraints from the age of Earth, the solar system, and at least some of the older globular clusters and extreme Population II stars.[12]

Other important neoclassical tests were source counts, redshift–magnitude relation, and redshift–angular size relationship. Source counts were actually deployed during the "great controversy" by Martin Ryle and his collaborators in pioneering radio astronomy, with the explicit aim of disproving the steady-state cosmological model. Although the test was not as decisive as Ryle wished it to be, it was certainly a key moment in the emergence of observational cosmology as we know it today.

The redshift–magnitude relation was established by Hubble, and he used its approximate version to demonstrate the eponymous law. It is an expression of the dimming of light from distant sources (notably galaxies, galaxy clusters, and in recent decades, supernovae) as it travels huge cosmological distances through space of a certain geometrical structure. Subsequently, many researchers, including Howard P. Robertson, Milton L. Humason, Nicholas U. Mayall, Allan Sandage, and Fredd Hoyle, have discussed the higher-order terms that, as expected, depend on the deceleration parameter q_0. This parameter is – at least in theory – a strong discriminant between cosmological models. In practice, it is subject to evolutionary uncertainties that led astray many observers who claimed to find various positive values for it, while we today know (after observing many supernovas of Type Ia at cosmological distances) that it has a negative value, $q_0 \approx -0.6$. Measurements getting the *sign* of the measured quantity wrong do not inspire much confidence! Of course, most of the uncertainties and problematic features of this test boil down to the uncertain evolution of stellar populations in galaxies over large spans of cosmic time, as well as to large uncertainties in correcting for extinction by dust, both at the source, in transit, and in the Milky Way. Due to these evolutionary uncertainties, galaxies (even first-ranked galaxies in clusters and other variations on that theme) are poor stand-ins for "standard candles."[13] Luckily for cosmology and its funding, supernovae of Type Ia proved much better.

The redshift–angular size test was first proposed by Hoyle in the middle of the great controversy as potentially the cleanest test of q_0, completely independent of extinction. The idea is to observe how the apparent angular size of distant sources, like giant elliptical galaxies or rich galaxy clusters, changes with redshift and to compare it with predictions of various world models. Unfortunately, the test replaces problems with "standard candles" with a host of problems with "standard rulers." Realistic astrophysical sources vary wildly by size, even in the local universe (at redshift $z \approx 0$), and the same sources tend to change their sizes with time. For instance, elliptical galaxies become larger by dynamical heating – supplying kinetic energy to "collisionless gas" of stars in tidal interactions with their galactic neighbors or even by cannibalizing smaller galaxies. Mergers of galaxies also occur much more frequently than hitherto suspected, and it gradually turned out that galaxies are surrounded by much more extended dark matter *and gaseous* haloes (or "coronae"), thus introducing another confusing element in the discussion of galaxy sizes.

On a more positive note, all those difficulties with neoclassical cosmological tests were a constant source of challenge and inspiration for observers, who kept trying to improve and hone them and to compare them with other observational data, notably those on primordial chemical abundances and the properties of the CMB. Confidence in the standard view of the universe at large scales is found only in tough-to-achieve synergy (or "concordance," a fashionable phrase in cosmological circles since the turn of the century). A historical perspective shows that it was right in the heat of the controversy before 1965 that most of these tests were devised and deployed, and some of the researchers who played a pivotal role in putting them forward, notably Sir Fred Hoyle, were among the dissenters from the mainstream view even after the epoch-making events of that year.

As we go on to show, the old-fashioned logical-positivist inspired view of the gradual and linear accumulation of knowledge is deficient. The two key ingredients of the modern cosmological worldview, namely the Hot Big Bang (and associated theories of nucleosynthesis, both primordial and stellar) on the one hand, and the new standard λCDM[14] model of structure formation and evolution (see Appendix A) on the other, emerged in dramatically different ways. Since they represent milestones in any meaningful discussion of cosmological knowledge, we summarize them briefly in Chapter 3.

3

Hot Big Bang and λCDM

Let us start with a thought experiment with time travel, of the kind popular among both theoretical physicists and philosophers. By turning the dial of an Herbert G. Wells's sort of time machine toward the cosmological past, we observe the evolution of the universe and all its parts in reverse. Distant galaxies are moving toward us, exhibiting systemic blueshift of spectral lines – while they themselves are becoming bluer (in the continuum) because of their younger and younger stellar populations. Stellar populations are becoming smaller, as more and more matter is in the form of metal-poor interstellar and intergalactic gas. At one point on our journey, galaxies themselves become dark, since there are no stars in them anymore, and subsequently dissolve into blobs of slightly increased density in a universal mixture of gas and dark matter particles. Matter and radiation in the universe get hotter and hotter as we go back toward the initial quantum state, because it is compressed into a smaller volume. As we approach the final singularity – which is, in reality, the initial singularity of the Big Bang – we experience the "early universe" conditions described by physical cosmologists. Our knowledge of physical laws and properties of various ingredients (baryons, photons, neutrinos, and to an extent, even unknown cold dark matter particles, gravitational waves, etc.) enables us to predict subsequent states of matter even in the very last second before singularity.

Returning now to the "real world" characterized by the cosmological arrow of time pointing out of the initial singularity, we can appreciate how the powers of physical *retrodiction* could have enabled George Gamow and subsequent cosmologists to create the Hot Big Bang scenario. Out of the initial singularity – or even pre-Big Bang epochs discussed within currently fashionable strands of quantum cosmology or string cosmology – matter expanded as a mixture of particles and radiation. Each ingredient had a particular epoch of decoupling, and some combined to form more complex structures when the decreasing temperature enabled corresponding bound systems to survive. Within the first second, many highly speculative processes, such as cosmological inflation and baryogenesis, took place; astronomers are seeking indirect confirmation of these in modern experiments, for example, by searching for relic primordial gravitational waves. At about $t = 200$ s, some baryons created light nuclei, such as deuterium, ^{3}He, ^{4}He, and ^{7}Li, as per the theory of primordial nucleosynthesis.

At about 300,000 years after $t = 0$, the start of the Hot Big Bang epoch, nuclei and electrons combined to form atoms. At earlier times when the temperature was higher, atoms could not exist, because the radiation then had so much energy, it disrupted any atoms that tried to form into their constituent parts (nuclei and electrons). Thus, at earlier times, matter was completely ionized and consisted of negatively charged electrons moving independently of positively charged atomic nuclei. Under these conditions, the free electrons interacted strongly with radiation by Thomson scattering. Consequently, matter and radiation were tightly coupled in equilibrium, and the universe was opaque to radiation. When the temperature dropped to the ionization temperature of about 4,000 K, neutral atoms formed from the nuclei and electrons, and this scattering ceased; the universe became mostly electrically neutral and very, very transparent. The fact that *observational cosmology exists* in the first place and is capable of seeing galaxies and their nuclear sources at enormous distances from us is a proof of that conjecture! The time when this transition took place is known as the *time of decoupling* – it was the time when matter and radiation ceased to be tightly coupled. The sea of radiation became what we today observe as CMB photons.

After decoupling, the process through which matter formed large-scale structures through gravitational instability, which started even earlier, was able to proceed much more efficiently. In particular, baryonic matter was able to fall into gravitational potential wells created by dark matter particles (which decoupled from the universal plasma much earlier); this, in turn, eventually led to the formation of the first generation of stars. The first stars aggregated matter through gravitational attraction, the matter heating up as it became more and more concentrated, until its temperature exceeded the thermonuclear ignition point, and nuclear reactions started burning hydrogen to form helium. Eventually, more complex nuclear reactions began in concentric spheres around the center, leading to a build-up of heavy elements (carbon, nitrogen, oxygen, for example), up to iron. These elements can form in stars because there is a long time available (millions of years) for the reactions to take place. Massive stars burn relatively rapidly and eventually run out of nuclear fuel. The star becomes unstable, and its core rapidly collapses because of gravitational attraction. If the stellar mass is above a certain threshold, the consequent rise in temperature and pressure blows it apart in a giant explosion, and new reactions take place that generate elements heavier than iron. This explosion is seen by us as a Type II supernova suddenly blazing in the sky, where previously there was just an ordinary star. Such explosions blow into space the heavy elements that have been accumulating in the star's interior, forming vast filaments of chemically enriched gas and dust around the remnant of the star (neutron star or black hole). This ejected material can later be accumulated during the formation of second-generation and all subsequent stars to form planetary systems around those stars – and eventually observers capable of formulating cosmological theories.

Since 1998 and the surprising discovery of dark energy and accelerated expansion of the universe, a new, improved standard cosmology has been emerging. It incorporates the standard Hot Big Bang cosmology and extends our understanding of the universe to times potentially as early as 10^{-32} s after the initial singularity, when the largest structures in the universe were still microscopic quantum fluctuations. This immediately implies that the new standard cannot be limited to tools of classical physics – extended with assorted borderline fields like nuclear physics and chemistry – but requires a full-fledged quantum field theory, including, at least in a conjectural manner, some aspects of the still non-existent quantum gravity (see Box 3.1).

Box 3.1

The *new standard cosmology* is characterized by

- A globally flat, accelerating universe.
- An early period of rapid, exponential expansion (cosmological inflation).
- Density inhomogeneities produced from quantum fluctuations during inflation, and hence about the same power at all length-scales (Harrison–Zel'dovich power spectrum of initial perturbation characterized as $P(k) \propto k^n$, $n = 1$; k being the wave number usually used instead of the characteristic wavelength λ – which itself should not be confused in this context with the dark energy amplitude).
- **Composition**: ~ 72% dark energy ($\Omega_\lambda \approx 0.72$); ~ 27% dark matter (both baryonic and non-baryonic); $0.5 - 1\%$ bright stars ($\Omega_m = 1 - \Omega_\lambda \approx 0.28$). Cosmological density fractions ("Omegas") refer to the ratio of density of any one component of the universe to the so-called critical density. Cosmological density of the component X is given as $\Omega_X \equiv \dfrac{\rho_X}{\rho_{\mathrm{crit}}} = \dfrac{8\pi G}{3H_0^2}\rho_X$, where G is the Newtonian gravitational constant, H_0 is the present-day measured value of the Hubble constant (expansion rate of the universe) and X could be any particular constituent of the universe (gas, dust, stars, Lyman-alpha absorbing matter, intergalactic medium, neutrinos, etc.)
- **Matter content**: ~ 23% cold dark matter; ~ 5% baryons; $\ll 1\%$ neutrinos and radiation ($\Omega_m = \Omega_{\mathrm{CDM}} + \Omega_b + \Omega_\lambda + \Omega_\nu$; $\Omega_{\mathrm{CDM}} \approx 0.23$, $\Omega_b \approx 0.05$, $\Omega_\gamma \sim 10^{-5}$, $\Omega_\nu \ll 0.01$)

This new standard cosmology, usually designated by the acronym λCDM, is certainly not as well established as the standard Hot Big Bang. (Note that while λCDM, coming from dark energy λ + Cold Dark Matter, in a strict sense is only the structure formation mechanism, the name is often used for the entire paradigm.) However, observational evidence is mounting from many sources. The most notable evidence is of the CMB power spectrum and anisotropies, but other important segments of modern-day observational cosmology include observations of supernovae at high redshifts, observations of first galaxies and quasars (QSOs – quasi-stellar objects), precision abundance measurements of various light nuclei like deuterium or lithium, and precision measurements of large-scale structures from the Sloan Digital Sky Survey (SDSS) and other galaxy surveys, today known as baryon acoustic oscillations (BAO; cf. Seo & Eisenstein, 2003).

Part II

Discovery of the CMB and current cosmological orthodoxy

4

Discovery of the CMB

The discovery of the CMB in 1965 by Arno Penzias and Robert Wilson became a turning point in the twentieth-century cosmology after it was interpreted by Robert H. Dicke and his coauthors. Penzias and Wilson were not looking for the CMB, which was, in fact, first predicted by George Gamow (1946) and his collaborators (Alpher, Bethe, & Gamow, 1948; Alpher & Herman, 1948a, 1948b, 1949) almost two decades earlier. Penzias and Wilson initially investigated the signal as noise, a possible artifact of their antenna or surroundings, and were completely unaware of its relevance to cosmology. The signal turned out to be a milestone in the development of physical cosmology. It ended the period of heated cosmological controversy (Kragh, 1996) and started a period of increasing convergence on the Hot Big Bang cosmology (Peebles, Page, & Partridge, 2009). It quickly became perceived as a standard but also initiated an interpretational race.

Gamow and his students (collaborators Robert Herman and Ralph Alpher) made an early prediction of the CMB temperature. Although their prediction was substantially off in terms of the value of its temperature, and turned out to be a calculating conundrum (Peebles, 2014, 210–211), it finally resulted in a value of 5 K (Alpher, Bethe, & Gamow, 1948). One the one hand, these papers pioneered the use of computing numerical methods in cosmology. On the other hand, it appears Gamow himself believed the radiation would be undetectable (Sharov & Novikov, 1993, 146).

It is easy to forget that these developments we see as a milestone today were the growing pains of cosmology, not a sudden revelation obvious to the entire physics community. Moreover, as Peebles noted:

The application of general relativity theory to the scales of cosmology was bold, because in the 1940s general relativity had passed just one serious test, the precession of the orbit of Mercury, and Gamow and colleagues were extrapolating the theory by 14 orders of magnitude in length from the orbit of Mercury to the Hubble length H_0^{-1}. (Peebles, 2014, 219)

The attempt was, to a great extent, a leap of faith.

Independent predictions by Andrei Doroshkevich and Igor D. Novikov (1964) and Dicke et al. (1965) were the result of much more thoroughly worked out models. In their 1965 seminal paper, Dicke and his collaborators suggested an interpretation of the CMB as a remnant of the primordial fireball, and this became the standard after Penzias and Wilson's discovery (Dicke et al., 1965). This interpretation postulated that in the early universe, baryonic matter and radiation were in balance. Baryonic matter is composed of protons and neutrons and also includes atomic nuclei; but in the very early stage (the first 200 s), it was only composed of free protons and neutrons. Thus, radiation consisted of continuously moving and scattering photons. This sort of balance, or equilibrium, of matter and radiation is similar to the one in human-made gas-discharge lamps; photons continuously "run around" and scatter off the matter. After the first light nuclei were formed about 200 s after the Big Bang, the universe was a hot, dense plasma composed of photons, electrons, and light ions, mainly protons and 4He nuclei. Now, the plasma was opaque to electromagnetic radiation – quite similar to fog – because of the strong Thomson scattering of light-photons by free electrons. Electrons were not yet tied to atomic structures but "floated" freely around. The mean free path each photon could travel before encountering an electron and scattering off it was very short, hence the fog-like opaqueness of the universe. Finally, a high ratio of photons compared to baryons (~ 109 times more of the former), starting with the early universe and extending to today, makes the Big Bang hot, that is, overwhelmingly heat- rather than matter-dominated (Figure 5.3).

As the universe was expanding and thus cooling, the formation of neutral hydrogen quickly became energetically favored. The fraction of free electrons and protons compared to neutral hydrogen decreased fairly rapidly, and the process was over about 400,000 years after the Big Bang. This process is labeled recombination, although electrons and ions met in the cosmological context for *the very first time*. As in other scientific fields, many key labels in astronomy are misnomers; the recombination was really the initial combination. In a way, such labels are devised to capture a particular view of a state of affairs. The term remains even though eventually the view changes. A more famous case of this sort that froze the key historical turning point in modern cosmology was the use of the label Big Bang. It was uttered as a joke by Fred Hoyle on a BBC show. As the key proponent of the steady-state universe, Hoyle devised a derogatory term that was meant to point out what he thought was a crucial weakness, even absurdity, of Georges Lamaître's expanding universe model that had a singular beginning in the form of sudden expansion. Yet over time, as expanding universe models became much more favored than static universe models, the term acquired an appeal because of the plastic description of the key point of the models, that is, the expansion from a singular moment in space and time.

Somewhat earlier, a process similar to recombination involving helium and an intermediate state of singly ionized helium Hep started; in this process, interactions of photons with neutral atoms replaced the Thomson scattering of photons by free electrons. The substitution is equivalent to switching the gas-discharge lamp off. The former interactions are several orders of magnitude less energetic than the latter, so interactions quickly subsided; photons effectively decoupled from (baryonic) matter and freely "flew" away. Their mean free path became comparable to Hubble length, or roughly the size of our cosmological domain, and the universe became transparent to light. This occurred about 400,000 years after the Big Bang. Photons from this epoch, or those incoming from the surface of the last scattering, traveled and cooled down freely, without interacting with matter, until some were stopped by a horn-shaped antenna in Holmdel, New Jersey, in 1964.

Given that these cosmological photons predate photons from almost all other sources in the universe, they show up as a background radiation "hum" – they ought to be detected as incoming from all directions in the sky. As the matter in the universe is isotropically distributed on large scales (i.e., equally throughout the cosmic space), many expected the primordial radiation must be distributed the same way. As we have seen, according to the Hot Big Bang model, the background radiation originated in the opaque state of plasma, so it should not be modulated, and it should thus behave as a perfect blackbody radiation (see Figure 5.1). This has been valid up to the order of minuscule anisotropies reflecting the beginnings of the physical structure formation. While such small-scale anisotropies are relevant in current cosmological debates, they were unobservable between the 1965 discovery and the advent of the *COBE* (Cosmic Background Explorer) satellite measurements in 1990s and, consequently, played no role in the overarching interpretation. As we will see, even the alternative interpretations focusing on the apparent isotropy of the CMB did not address them; they either bet on deviation from the blackbody radiation shape or modified the hot beginning of the Big Bang by devising plausible ways of getting current isotropy from early fluctuations.

It was also immediately clear within this framework that the original temperature of the blackbody radiation during recombination was in thousands of kelvin and should have steadily decreased due to the expansion of the universe to a few, or perhaps a few dozen, kelvin. This parameter was arguably crucial in testing the model, solidifying the interpretation of the background radiation within it, and eliminating the alternatives.

As we mentioned earlier, George Gamow and his students Ralph Alpher and Robert Herman developed a relativistic Hot Big Bang model that also predicted a primordial relic radiation some 20 years before the discovery by Penzias and

Wilson. And in fact, right after the publication of Lemaître's (1931) speculative column in *Nature* on the possibility of the creation of the universe from a single quantum, the physicists realized any such scenario would result in leftover radiation of some sort, and this was commonly understood as a good reason not to accept such models (Gregory, 2005, 38). The account was ignored, and with it, Gamow et al.'s understanding that a very hot initial state must have been opaque to light and that the subsequent recombination would cause photon decoupling and the emergence of a cosmological photon reservoir with adiabatically decreasing temperature. A paper arguing this point was published in 1949 (Gamow, 1949), but these insights dated back to 1946. And Alpher and Herman (1948a, 1948b, 1949) predicted that the temperature of the background is about 5 K. Later on, this prediction became known as the "temperature of the universe."

It is not surprising that Gamow and Alpher and Herman kept revising the value, as their procedures for determining it were inherently problematic. The two students used the equilibrium Saha approximation, but it included a poorly estimated range, between 5 K and 50 K. Gamow used the Jeans stability criterion, which was unfortunately inapplicable in the simple form he used due to the presence of dark matter he did not know about. Generally speaking, all three pictured the transition from the radiation-dominated to the matter-dominated universe as occurring far too late and at far too low an equilibrium temperature, at least based on current insights, due to their assumption of the baryon density of the universe that was too high – it turned out much later that those baryons comprise only about 5% of the total mass-energy budget.

Their model and their prediction were well worked out given the available knowledge at the time, but they were not taken seriously until the emergence of the great controversy forced astronomers to search for observational tests of the models. One key reason why their model did not take off earlier within the community was that the model's non-equilibrium calculations required a numerical solution of differential equations that could be realistically done only on a computer. Computer calculations were still uncommon in the 1940s and 1950s when they were developing their model. The required computer calculations were performed independently by Dicke and Jim Peebles at Princeton and Doroshkevich and Novikov in Moscow between 1960 and 1964 (Doroshkevich & Novikov, 1964; Dicke et al., 1965). The initially obtained values were still too high, at about 40 K, and were reduced ten-fold with Penzias' and Wilson's discovery.

In an ironic twist of events, the possibility that a relevant signal of the background radiation was detected much earlier was raised by none other than Sir Fred Hoyle, the key proponent of steady-state cosmology. He stated the CMB had indirectly been observed as far back as 1941 by Andrew McKellar, a Canadian astronomer working at the Dominion Astrophysical Observatory in British Columbia, although he took it as a refutation of Gamow et al.'s proposal, taking into account the 50 K prediction in Gamow's book, instead of the 5 K prediction of Alpher and Robert Herman (Peebles, 2014, 220). McKellar (1941) observed the rotational excitation of the cyanogen (CN) molecules toward the star z Ophiuchus. The data clearly indicated that some external factor was exciting those rotational transitions. The subtraction of energy calculated from local excitation indicated excess energy and the corresponding temperature of the excess source. McKellar assumed it was a contact with the thermal bath and, based on this, estimated the "temperature of deep space" to be about 2.3 K. Contemporary studies of the same sort have confirmed the uncanny coincidence of this estimate with the modern value for the CMB temperature (e.g., Roth, Meyer, & Hawkins, 1993). These studies have also found that the CMB indeed excites the lowest rotational levels of interstellar cyanogen across interstellar clouds. In fact, these sorts of measurements are another important line of evidence for the origin of the CMB and the prediction of the CMB temperature at earlier times. And as we will see, they played a role in defending both the orthodoxy and the alternatives. The temperature of the CMB should increase in a linear fashion as we observe more into the past, and the temperature of distant gas clouds should be detected in accord with linear increase, once the local kinetic energy is known and subtracted. In 2000, the spectrograph mounted at the European Southern Observatory (ESO) measured the temperature of such a cloud at $z = 2.3371$ (Srianand, Petitjean, & Ledoux, 2000). The result was in excellent agreement with the *COBE* results and served as an independent source of evidence for the CMB's temperature.

The first deliberate testing related to the model was performed in Florida in 1946 (Dicke et al., 1946). It tested the prediction of Dicke and collaborators that at a temperature less than 20 K, the CMB would not show much leftover radiation at radiometer wavelengths.

All these tests and discoveries were merely the first segment in the long road to acceptance of the Hot Big Bang model. It was opportune that a sufficiently sophisticated theoretical framework already existed when the background signal was undeniably discovered. This contributed to a fairly quick and wide perception of the model as an orthodox interpretation, making it a target of alternative accounts. Over the following several decades, this interpretation gradually became an integral part of the standard cosmological paradigm, to such an extent that abandoning it now would amount to a wholesale rejection of the entire edifice of physical cosmology.

5

CMB phenomenology

Before we dive into the rich and diverse world of explanations of the CMB follow-ing Penzias and Wilson's discovery, we will present the CMB and its properties from the standpoint of the now overwhelmingly accepted Hot Big Bang theory. The extraordinarily successful *COBE* mission provided a plethora of relevant data that resulted in decades of numerous observational and theoretical studies of the CMB. The satellite realized three objectives: detailed testing of the shape of the blackbody spectrum, mapping the CMB across the sky, and detecting diffuse back-ground in infrared and millimeter range (possibly resolving early cosmic objects, stars, and galaxies). The goals of the mission were, in fact, a result of decades-long debates on the nature of the CMB and its basic properties (see Figure 5.3).

Following the observational studies of the acquired data, the first result concern-ing the first key property of the CMB essential to any attempt to explain its nature was that the background radiation is a blackbody to very high precision levels (Mather et al., 1994; see Figure 5.1). Moreover, its temperature, the second key property, has been measured with unprecedented precision. According to the *COBE* dataset, it is 2.728 ± 0.004 K with a 95% confidence level (Fixsen et al., 1996). A more recent *COBE* dataset with *WMAP* (Wilkinson Microwave Anisotropy Probe) recalibration suggests the value of 2.72548 ± 0.00057 K K (Fixsen, 2009), while an even more recent *WMAP* and *Planck* dataset (Hinshaw et al., 2013), which is still being improved, suggests 2.7260 ± 0.0013. All these results are quoted in the standard $T \pm \Delta T$ format, from which the relative amplitude of anisotropies $\Delta T / T$ is readily obtained.[15]

We should note that one of the most impressive recent results is the measure-ment of the CMB temperature by the European Southern Observatory (ESO) at the epoch corresponding to z of approximately 2.34, as mentioned earlier, using properties of the molecular hydrogen in a damped Ly-alpha absorption system in the spectrum of background QSO (Srianand, Petitjean, & Ledoux, 2000). The obtained result, although characterized by a large error margin, corroborates the predictions of the standard CMB interpretation.

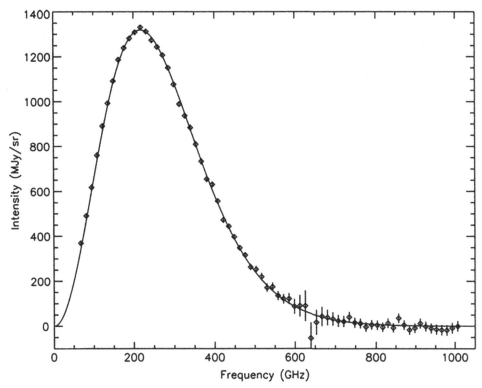

Figure 5.1   Blackbody spectrum of CMB as established by the *FIRAS* experiment on board *WMAP*. From Fixsen (2009). (The copyright obtained from the author and the publisher, ACS.)

Finally, precision measurements of isotropy, the third key property of the CMB, have been obtained. Thus, the CMB is fairly uniform over the sky. Except for the already detected dipole anisotropy due to the motion of the observer together with the local galactic group (see Appendix B), the abovementioned very small anisotropies have been detected only recently at large angular scales (7° and larger) and at the extremely faint level of $\Delta T / T \sim 10^{-5}$ (Smoot et al., 1992; Hinshaw et al., 2003, 2013; see Figures 5.2 and 5.3).

These three crucial properties – the shape of the spectrum, temperature, and isotropy – are not completely independent. We could not speak of the global CMB temperature were it not for its blackbody spectral shape and unusual isotropy. In fact, if we follow the standard interpretation, the spectral shape and isotropy are the primary properties of the CMB, while the temperature is an epiphenomenal property that, theoretically speaking, is adjustable, contingent on other nonessential factors. In one of the particularly serendipitous moments of modern history of cosmology, three decades before the discovery of the CMB, Richard Tolman (1934),

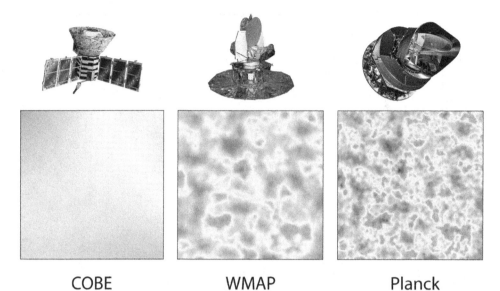

COBE   WMAP   Planck

Figure 5.2   The advancement of space-based CMB observatories: While *COBE* discovered intrinsic anisotropies in the CMB (those which are not consequences of our motion), the *WMAP* and *Planck* missions obtained insights into the map of the CMB. Credit: By NASA/JPL-Caltech/ESA – http://photojournal.jpl.nasa.gov/catalog/PIA16874 (direct link), Public Domain.

Figure 5.3   *WMAP* all sky survey of CMB anisotropies. The Internal Linear Combination Map minimizes the Galactic foreground contribution to the sky signal. It provides a low contamination image of the CMB anisotropy, which translates into the angular-scale power spectrum of primordial inhomogeneities. It is, arguably, the major tool of contemporary cosmologists. Credit: NASA/*WMAP* Science Team.

in his influential book, proved that the Hubble expansion of the universe would preserve the blackbody shape of any initially present blackbody radiation, with only the temperature decreasing linearly with the scaling factor. As we will see in due course, this physical fact makes the standard CMB interpretation seem "natural" but interferes with some of the attempts to interpret the radiation as a patchwork of sources thermalized at different epochs.

A significant and frequently cited consequence of the orthodox interpretation of the CMB that figured prominently as a target of the alternatives is the limit the temperature of the background radiation sets on the fraction of the universal density in the form of baryonic matter. The simplicity of the physical view that results in a predicted limit is part of the appeal of the standard model. Thus, if we leave aside negligible, very slow, physical processes affecting particles, like potential proton decay, and approximate at timescales comparable to the Hubble time, the baryonic number turns out to be a conserved quantity, while the vast majority of photons currently existing in the universe are the photon relics detected as the CMB. Therefore, the photon-to-baryon ratio today is essentially the same as it was at the time of decoupling – a remarkably simple trait of the universe. Finally, if the limitations on the baryon-to-photon ratio in the early universe, which are based on the theory of primordial nucleosynthesis (Copi, Schramm, & Turner, 1995; Schramm & Turner, 1998), are coupled with the fixing of the photon density per co-moving volume, we obtain a unique handle on the total cosmological baryon density Ω_b.

As we have pointed out, this value played a major role in the formulation and defense of the alternatives. We will look at the details of these arguments, but we should note that it is obvious that this value could have taken cosmologists in various directions. And already in 1968, John R. Shakeshaft and Webster (1968) demonstrated that the energy density ratio of primordial to non-primordial radiation is about 400 to 1. They drew this conclusion independently of the interpretation of the CMB. Moreover, even in the steady-state universe, the total number of photons emitted by conventional sources, such as stars within a sufficiently large co-moving volume, diverges. So does the number of thermalized photons originating with a hypothetical early stellar population (usually called Population III, although the term is occasionally used to denote any object with primordial chemical composition or zero metallicity), which provided both the first metals and the energy of the CMB. It is not surprising, then, that Sir Fred Hoyle repeatedly used this "coincidence" to argue for the Population III origin of the CMB (e.g., Hoyle, 1994).

6

Standard "textbook" history and its shortcomings

As we write these words, although some doubts are voiced from time to time (e.g., Baryshev, Raikov, & Tron, 1996), the astrophysics community has pretty much converged on the orthodox interpretation of the CMB. Moreover, the standard cosmological model is firmly founded on both available evidence and theoretical studies. And the 1965 discovery was clearly a watershed moment for both of these developments. Its magnitude makes it really difficult to resist the view that the now-standard interpretation of the *discovery* of the CMB, as a discovery of a remnant of primordial fireball, was an inevitable, fully transparent, and even predictable moment. Without dwelling on the details of forgotten history, it is very hard to fathom that any alternative interpretations would have been offered, seriously or even semi-seriously, by any distinguished cosmologist at the time. As we generally prefer to focus on and remember a triumphant scientific moment rather than ponder the winding and rocky road that led to it, with all its tributaries and blind alleys, the impressions of the inevitability of the current view and the shiny moment of 1965 are widely shared by astronomers and laypersons alike.

In fact, the textbooks reinforce such a view. Two of the best cosmology textbooks available, by Peter Coles and Francesco Lucchin (1995) and John Peacock (1999), are great examples of such reinforcement. For example, Peacock formulated a mind-bogglingly simplistic, if poetic, statement about the CMB: "The fact that the properties of the last-scattering surface are almost independent of all the unknowns in cosmology is immensely satisfying and gives us at least one relatively solid piece of ground to act as a base in exploring the trackless swamp of cosmology" (Peacock, 1999, 290). Using a less flourishing style and adding a dose of reflection on the history of the field, in his commentary on the re-edition of Penzias and Wilson's (1965) paper, Peebles wrote:

A willingness to believe such an elegant gift from nature surely also played a significant role in the early acceptance of the CBR [cosmic background radiation] interpretation ... During four decades of involvement with this subject, I have grown used to hearing that such advances have at last made cosmology an active physical science. I tend to react badly because I think cosmology has been an active physical science since 1930, when people had assembled a set of measurements, a viable theoretical interpretation, and a collection of open issues that drove further research. This equally well describes cosmology today. (Peebles, 1999, 1067)

Albeit unwittingly, the passage sets the stage for a deeper philosophical-historical analysis by pointing out two key factors. First, the "willingness to believe" the standard model was certainly a major motivator of the push to develop the orthodoxy and was also the reason for a subsequent lack of reflection on the roads that led to it. Second, the discovery of the CMB led to a subsequent lack of confidence in the pre-1965 cosmological research, and this significantly contributed to a streamlined view of the history of physical cosmology (Peebles, 2014). These two factors merged into a widespread impression that the microwave noise detected serendipitously by Penzias and Wilson *suddenly* threw us into an epoch of serious, quantitative cosmology, and the essential validity of the Hot Big Bang paradigm occurred pretty much *instantly* and has remained unchallenged ever since. The following sentence by Coles and Lucchin summarizes this attitude: "it is reasonable to regard this discovery as marking the beginning of 'Physical Cosmology'" (Coles & Lucchin, 1995, xiii).

There is a dearth of good historical studies on the recent stages of our cosmological adventure, and there are multiple reasons for this. One reason touches on a long-standing controversy about what has been called "the Whig interpretation of history": the tendency to portray the past as the inexorable march of progress toward the present enlightened and desirable state. The crux of the matter is Herbert Butterfield's injunction:

Real historical understanding is not achieved by the subordination of the past to the present, but rather by our making the past our present and attempting to see life with the eyes of another century than our own. (Butterfield, [1931] 1959, 16)

"Making the past our present" has become a slogan of anti-Whig mainstream historians. A useful analogy would be travel in space where those vistas and customs most distant from our own are the key object of interest for a traveler; it would be a poor traveler who would denounce the denizens of distant regions for the dissimilarity of their customs. Similarly, Butterfield invited us to regard historical research as something similar to traveling to another time with the help of a Wellsian time machine, without prejudicing or denouncing the denizens of those times. This position is uncontested in political or economic history, as well as history of art; perhaps the only field where it is seriously doubted or even rejected is history of science, since there is no reasonable construal on which science has not made objective and often measurable progress (e.g., Hall, 1983; Harrison, 1987; Mayr, 1990; Jardine, 2003; Alvargonzález, 2013; Oreskes, 2013; Giunta, 2022).

A couple of pages later, Butterfield stated another point very salient for our present purposes: "The things which are most alien to ourselves are the very object of his [historian's] exposition" (Butterfield, [1931] 1959, 18). Unexplored – or only weakly explored – alternative explanatory hypotheses agree with this goal. So in a sense, by doing history of recent science, we strike a narrow compromise, something mentioned by the distinguished astronomer Edward Harrison in his well-known critique of Butterfield's approach:

Without some compromise between historians and scientists – between anti-Whiggery and Whiggery – recent history of science must suffer from lack of professional attention. An anti-Whig historian who knows little modern science will skip this period and explore earlier and much safer periods; a scientist who fears the opprobrium associated with the Whig epithet will also avoid this period and write only reviews of current developments. ... Better to have eye-witness accounts than historians' guesses centuries later. (Harrison, 1987, 214)

This is indeed a serious problem, not significantly discussed since the publication of Harrison's paper. Our decision in the present book is to entirely ignore "the opprobrium associated with the Whig epithet" and to employ various methods of historical and philosophical research, disregarding more or less fashionable labels and derisions. In studying a relatively novel phenomenon like the CMB (from the point of view of human science, it is not intrinsically novel), there is simply no escape from keeping an eye on the standard interpretation and powerful theoretical framework of modern physical cosmology. To do otherwise would be a dereliction of duty to the prospective audience that, similar to soccer audiences, is fully entitled to know the score of the match. Since everyone is likely to concur that history of science is not written just for historians of science – especially not just those historians of science steeped in the non-Whiggish gospel – this duty should be taken seriously.

We could go a bit farther and suggest the very definition of the topic is impossible under the strict construal of non-Whiggish history of science. Harrison argued from a cost-benefit analytic viewpoint ("better to have eye-witness accounts, etc."), but we could adopt a stricter stance and ask, "Eye-witness accounts of what exactly?" The boundary of a topic or a subfield cannot be properly set following the strictures of non-Whiggism, not only since the particulars of our scientific knowledge, or even just scientific speculation about a particular phenomenon are biased by our present-day dominant scientific paradigms, but also because the very existence of the phenomena under scrutiny is often part and parcel of those dominant paradigms. Since history of science, as any historical discipline, depends on the process of selecting available evidence on the basis of relevance (as well as some further criteria), this process cannot even begin without present-day scientific knowledge, alleged inadmissible by the non-Whiggish fundamentalists.

Nowhere is this seen better than in the case of the CMB origin and properties. Sir Fred Hoyle brought the attention of astronomical audiences to Andrew McKellar's 1941 rather obscure publication on the "rotational temperature" of interstellar space as a precursor to the discovery of the CMB, but his act could only be understood when the importance of the CMB photons was realized, and that came later. There is no way to properly assess the relevance of Hoyle's act – notwithstanding his motivations – without both the factual input about the "official" discovery of the CMB in 1965 and the background understanding of the crucial importance of this discovery in the context of cosmological controversies.

We give many such examples throughout this book. For that reason, we regard the book as a contribution to the perennial debate on the nature of historical knowledge and historical analysis, especially Whiggish versus non-Whiggish histories.

Our central goal, then, is to understand the emergence of the consensus, the exact role alternatives and their failures played in its formation, and the epistemological attitudes that drove all this. We develop our historical account of the discovery of the CMB and the great controversy that preceded it with that central aim in mind. In effect, we provide the look probably at the first prolonged debate, a fruitful scientific controversy, so to speak, documented according to the current standards of currently desired historical detail. The exhaustive publication practices of today had just caught on around the time of the controversy's emergence, so every theoretical and empirical detail and idea is well documented and can be put into a larger historical context.

Yet we are not offering a comprehensive history of the discovery of the CMB. The historical accounts the reader may want to consult are those by John North (1994) and Bruce Partridge (1995). A history of measurements of the CMB temperature up to the early 1980s appears in Chapter 12 of Jayant V. Narlikar's 1983 work. General accounts of modern cosmological paradigm can be found in many advanced textbooks, but we most frequently consulted Peebles (1993).[16]

Having stated our caveats, we can now summarize our account of the gradual emergence of the orthodoxy. There were four distinct stages on the road to the full-scale establishment of the orthodox interpretation of the CMB. First, the theoretical model of the Hot Big Bang was developed, starting in the 1940s. Its various implications and parameters were, at least in principle, observationally testable, although it took decades before astronomers were convinced of realistic chances of such tests. Second, in the 1960s, independent groups of astronomers and physicists performed crucial calculations very precisely with the use of computers of the model's empirical implications, including the properties of the background relic radiation. Third, the discovery by Penzias and Wilson in 1965 was quickly and increasingly seen as a successful test of the Hot Big Bang model. At this stage, a moderate convergence of agreement on the model and the interpretation of the discovery within its framework emerged quickly yet left a sizeable domain for the development of various alternative explanations over the following two and a half decades. The dynamics of the development of the alternatives was sensitive to the surprisingly varying results, first on the shape of the CMB spectrum (especially in the late 1970s and early 1980s) and its apparent deviations from the shape of the blackbody spectrum, and second on the measurement of its isotropy. The number and the variety of the results during that period were such that in a 2009 publication of the *WMAP* milestone results, Dale J. Fixsen noted, with a tinge of after-the-fact superiority, "[t]here were many publications of measurements of the CMB temperature from the late '60s and '70s, but the uncertainties are large and the systemics were not well understood." Thus, they could be excluded from the combined values of more recent and more precise measurements he then presented. The fourth stage, in the early 1990s, was marked by the COBE satellite observations and their unprecedented precision. The results were quickly and overwhelmingly interpreted as truly identifying the origin of the CMB, thereby cementing a very wide convergence of agreement on the Hot Big Bang model and the interpretation of the CMB within it. Further "alternative research" has quickly become difficult, if not impossible in practical terms. Certainly no institutional support of appreciable direct funding has been forthcoming.[17]

As aspects of the CMB were discovered over the next several decades, the postulation of alternatives became increasingly constrained, especially after the *COBE* discoveries. Yet between the initial, moderate convergence and the later wide convergence some two and a half decades later, a number of alternatives were developed, defended, criticized, and repositioned. These developments were often reactions to the key content developed in the four crucial stages of the emerging orthodoxy or prompted by incoming data. Thus, to explicate and assess the alternatives, we will identify the crucial aspects of the orthodoxy as they relate to the understanding of the CMB in each instance.

7

Emergence of precision cosmology

Given the constant stream of observations relevant to cosmology provided by big science astrophysics, namely the Hubble Space Telescope and its recent successor the James Webb Space Telescope, we may perhaps overlook the fact that cosmology is the youngest of all physical sciences. Like other subdisciplines of physics, astronomy has experienced bouts of big science across centuries (e.g., Tycho Brahe's Hven Island project, old Mongolian, Indian, and Mayan observatories, and possibly Stonehenge). Physical cosmology, however, entered the family of observational sciences in a substantial way in the 1960s with the discovery of quasars by Maarten Schmidt and with Penzias and Wilson; arguably, it fully entered the big science network only in the late 1980s and early 1990s. Until then, it was a field developed in small, often interdisciplinary circles composed of intellectually ambitious scientists ready to swim against the mainstream by applying the latest achievements of physics to daring cosmological questions.[18]

In their ground-breaking report, Penzias and Wilson provided the values of the key observational parameters and prior expectations of the values and carefully drew a potential theoretical-observational inference:

Measurements of the effective zenith noise temperature of the 20-foot horn-reflector antenna (Crawford, Hogg, & Hunt, 1961) at the Crawford Hill Laboratory, Holmdel, New Jersey, at 4080 Mc/s have yielded a value of about 3.5 K higher than expected. This excess temperature is, within the limits of our observations, isotropic, unpolarized, and free from seasonal variations (July 1964–April 1965). A possible explanation for the observed excess noise temperature is the one given by Dicke et al. (1965) in a companion letter in this issue. (Penzias, & Wilson, 1965, 420)

It is indicative of the things to come that Robert H. Dicke and coauthors provided the first discussion of alternative interpretations of their model. After an elaboration of the Hot Big Bang model emphasizing the presumed isotropy and uniformity of the universe, they explained why an oscillating universe alternative or open universe cannot be accepted, given the temperature of 3.5 K; they developed a fairly elaborate account of the possibilities of a closed universe and its problems, concluding that the required leptons-to-baryons ratio is too high (Dicke et al., 1965, 418–419).

If the discovery by Penzias and Wilson was a key step for cosmology as an observational science using regular astrophysical methods of observation, thus ceasing to rely exclusively on abstract mathematical or physical modeling, then its first step into the territory of big science was the experiment with the CMB performed in 1977 on a U-2 military aircraft by George Smoot, Marc Gorenstein, and Richard Muller (1977). The apparatus (Figure 7.1) was calibrated and collected data over eight flights of the aircraft. First, the atmospheric background and anisotropy due to the motion of the earth were eliminated by measurements with a twin antenna, while spurious anisotropies were eliminated by the aircraft changing flight direction every 20 minutes. The anisotropy due to the Doppler shift was expected and finally measured at 33 GHz: "The cosine anisotropy is most readily interpreted as being due to the motion of the earth relative to the rest frame of the cosmic blackbody radiation" (Smoot, Gorenstein, & Muller, 1977, 899). The researchers also detected a surprising velocity of the Milky Way and the local group of galaxies (Smoot, Gorenstein, & Muller, 1977, 900).

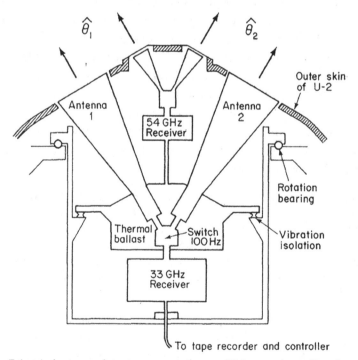

Figure 7.1   Anisotropy detector mounted on a U-2 spy plane. Reprinted with permission from Smoot, G. F., Gorenstein, M. V., & Muller, R. A. (1977). Copyright (1977) by the American Physical Society.

We will keep returning to these ground-breaking results in the study of the CMB. For now, we note how striking it is that the authors of the experimental report were open to the alternative interpretations of discovered cosine anisotropy as possibly inherent to the CMB, even in 1977. They wrote: "We cannot eliminate the possibility that some of the anisotropy is due to an intrinsic variation of the cosmic blackbody radiation itself" (Smoot, Gorenstein, & Muller, 1977, 901). While they were probably aware of the seminal paper of Rainer Sachs and Arthur Wolfe (1967) proposing the mechanism for generating small-scale CMB anisotropies from early overdensities – thus linking the temperature map with the process of structure formation and evolution – it is doubtful that they recognized its ramifications. Citation analysis shows Sachs and Wolfe garnered about dozen or so citations per year for the first 25 years (!) after publication, but this exploded by about tenfold in 1992, after *COBE*.

Another two measurements turned out to be milestones. One was performed by David P. Woody and Paul L. Richards (1978, 1979) from a hot air balloon and first published in 1978 as a preprint, and the other one by the *COBE* mission and published by Mather et al. in 1990. The measurement by Woody and Richards was initially corroborated by Herbert P. Gush's 1981 measurement from a rocket (Gush, 1981) and then by a collaborative measurement involving Richards from a rocket a few years before the *COBE* mission (Matsumoto et al., 1988). Mather et al. (1990) provided the first comprehensive *COBE* results, using essentially the same apparatus, but this time mounted on a satellite. This second string of measurements illustrates the progress in measurement techniques, the power of the results of measurements to sway theoretical attitudes, and the slow convergence process with all the twists and turns that were later forgotten, or at best diluted and distorted to create the appearance of smooth sailing.

Woody and Richards used a spectrophotometer mounted on a balloon and cooled it by liquid helium. Their results showed substantial deviation of the CMB spectrum from the Planck curve, from 10% to 20%, depending on the area of the spectrum. The measurements were performed at various zenith angles and floating pressures to avoid artifacts, and the authors estimated the confidence level at 85% (Woody & Richards, 1978, 6; Figure 7.2). A number of the alternatives to the Hot Big Bang model were invigorated with these results, as they were the most precise at the time. We will encounter this observational milestone across different categories of the alternative interpretations.

It should be noted that this was not the first of this sort of measurement. The ground measurements were exposed to thermal conditions of the lower atmosphere that were much warmer than the signal they were seeking. One way around this impediment was to repeatedly switch between pointing antennae at the cooled helium bath and at the sky. Yet the key problem was that the upper

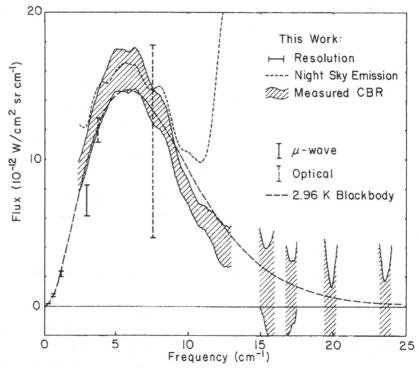

Figure 7.2   The spectrum of the CMB obtained with a spectrograph mounted on a balloon in 1979, exhibiting deviations from the shape of the blackbody spectrum. Only the advent of the *COBE* mission unequivocally eliminated this deviation. Reprinted with permission from Woody, D. P., & Richards, P. L. (1979). (Republished with the permission of one of the authors and the publisher.)

layers of atmosphere interfere precisely in the part of the spectrum (less than 1 cm in wavelength) that is crucial to measure to see whether the CMB fits the shape of the blackbody spectrum, or whether it complies only partially (following the Rayleigh–Jeans law capturing wavelengths up to the microwaves, but not Planck's law). Thus, in the late 1960s and early 1970s, several ground, balloon, and rocket measurements were taken, but the results varied substantially. The proponents of the orthodoxy expected divergences from the blackbody spectrum shape to coincide with the "unexpected sources of radiation at high altitude" (Weinberg, 1972, 515), while those keen on developing alternatives found this improbable. These uncertainties could be resolved only with infrared measurements from a satellite, outside the atmospheric and other thermal interferences, by cooling the equipment. In 1978, Woody and Richards introduced the cooling of the equipment, thus adding to the credibility of their results.

Fast forward 22 years: the same measurement technique, with essentially the same detector Woody and Richards used, but this time mounted on a satellite

(COBE) devoid of atmospheric interferences, removed any reasonable doubt that the CMB spectrum was deviating from the blackbody curve by any value larger than 1%. The authors, in fact, cited the work of Woody and Richards when describing their own equipment, pointing out similarities, but also mentioning "several improvements" (Mather et al., 1990, L37). Moreover, the Far InfraRed Absolute Spectrophotometer (*FIRAS*) mounted on *COBE* addressed the possibility of unresolved sources – the key hypothesis for a few alternative interpretations – and refuted it with unprecedented precision. We will be returning to this epoch-making measurement throughout the book as well. In between these two milestone results, George Smoot and colleagues' (Smoot et al., 1987) measurements – part of a large international collaboration – reduced the divergence from the blackbody shape of the spectrum to less than 6%.

Part III

What constitutes an unorthodoxy? An epistemological framework of cosmology

8

Underdetermination of theories
and models in cosmology

Studying the emergence of alternatives to a standard theoretical account of a physical phenomenon can tell us a lot about the scientific process, the way scientific knowledge is produced, and the way key discoveries are made. There is certainly a sociological aspect to this, as Martín López-Corredoira (2014) convincingly argued, focusing on alternative cosmological models. We are cognizant of this particular aspect, although our primary interest is methodological and epistemological. More specifically, we examine the interplay of theory and evidence in fashioning the orthodoxy and its alternatives. Our socio-epistemological treatment of relevant social aspects of this interplay is an organic part of our analysis. In particular, we ask how certain epistemic attitudes of researchers motivated and shaped their theoretical pursuits and how the entry of cosmology into the big science club changed the nature of the entire enterprise. This question has been on the minds of practicing physicists and cosmologists for some time. The fear that this growth may have tangible consequences was dramatically expressed by Halton Arp, whose work on supposed association of quasars and nearby galaxies became controversial: "One danger today ... is that with science tied increasingly closely to expensive equipment, which in turn is tied to prominent institutions, ... science may progress less rapidly than its potential" (Arp, 1987, 169). Fred Hoyle extended public support for Arp, if not for his theory (Gregory, 2005, 330). Hoyle warned of the danger of organizing scientific institutions that tend to prematurely kill novel ideas just because they do not comply with the mainstream (Gregory, 2005, 331). In fact, because Hoyle and Arp swam against the current, along with John Maddox and Geoffrey Burbidge, the journal *Nature* introduced the section Hypothesis, discussing ideas outside the mainstream but worthy of consideration (Gregory, 2005, 333).

Philosophical and methodological discussions of the emergence of the alternatives and implications for understanding scientific method and its epistemological ramifications have focused on a range of major episodes and achievements in physical sciences. According to Cushing (1994), Bohm's mechanics, a viable and in few important respects superior alternative to the dominant Copenhagen interpretation of quantum mechanics, was sidelined initially because it was devised too late. Thus, a historical contingency led to widespread acceptance of the early orthodoxy (Cushing, 1994). A counterfactual historical analysis was developed to back this view up (Soler, 2023). In chemistry (Chang, 2009, 2010), now forgotten abandoned alternatives were, at the time, viable theories revealing alternate routes to the discoveries that had been made; these viable routes were never taken far enough because of the contingencies of history. Subtle and seemingly insignificant changes in experimental conditions in modern particle physics prevented the earlier emergence of crucial theoretical accounts (Perović, 2011). In a case unfolding in current fundamental physics, Dawid, Hartmann, and Sprenger (2015) offer a Bayesian analysis of theoretical preferences when experimentally viable theoretical alternatives in fundamental physics were not available. And Kyle Stanford (2006) offered comprehensive examples of unconceived alternatives in modern biology.

The production of alternative theories and models seems especially open-ended in historical scientific fields, such as geology, paleontology, and archaeology. These fields concern events very deep in the history of nature or society, so they are bound to observationally reconstruct limited samples of potential evidence. This typically leaves a wide space for the construction of alternative interpretations. In addition, as Ellis (2014, 2) pointed out, cosmology is unique among the historical sciences because the subject matter of its study is unique, namely a unique universe, while other such sciences typically benefit from cross comparisons of similar deep past phenomena of interest.[19] We might expect that the epistemic standing of orthodox thought in such fields is inherently tied to the epistemic standing of available alternatives, and these are often easier to devise in these fields than in the sciences without such limitations. The battle of competing theoretical accounts in light of the available evidence is bound to be pronounced and long lasting – the evidence needed to break the stalemate is often hard or impossible to come by, and a crucial, controlled experiment varying key parameters treated differently by the competing account cannot be devised, even in principle, when the phenomenon at stake is a one-off social, planetary, or cosmic event. In contrast, at any given moment, barring the sometimes-detrimental socioeconomic constraints, experimental physics can provide tangible experimental grounds for theoretical accounts in the form of direct evidence addressing several parameters at stake by varying and controlling them. This quickly and severely constrains theoretical accounts of relevant phenomena.[20]

Philosophers of science have devised the notion of *underdetermination* of theories by evidence in order to capture situations where the evidential basis does not unequivocally decide between different theoretical accounts. This notion is useful for our analysis as it captures the essence of the methodological and epistemic situations during the development of cosmology, as well as the points made on this situation by its proponents.

In simple terms, the underdetermination thesis, as we use it, asserts that our observations may be insufficient to definitively determine which one of multiple competing hypotheses or theories explaining a physical phenomenon is superior. Moreover, the accumulation of more precise observations might not necessarily result in a clear winner among these competing theories either. Eventually, a preponderance of evidence emerges in favor of one theory, as recognized by the vast majority. Thus, underdetermination of theories by evidence in this case is transitory, but crucially, it can be prolonged as well, allowing for plausible alternative viewpoints (challengers to the established perspective) to emerge based on the available observations. In such a case, the ultimate convergence toward a single theory does not occur abruptly or swiftly; rather, it is a gradual, complex process. The conventional account of the history of the CMB discovery and its aftermath overlook precisely such an extended period, prematurely asserting the conclusiveness of theory selection immediately after 1965. In our discussion, we delve into the reasons behind the persistent indecisiveness in the protracted underdetermination of competing hypotheses regarding the origin of the CMB.

Yet we will use the notion cautiously. We need to qualify and clarify the way we will use it to prevent it becoming an impediment, especially because the notion has been used in the literature under very different assumptions. Our clarification goes to the heart of the epistemic and methodological playing field of modern cosmology and its historical development.

A highly abstract notion of underdetermination of theories by evidence, initially perpetuated in the early works of Willard V. O. Quine (1975), is supposed to grasp inherent, epistemic uncertainty of the sort we customarily encounter in all scientific fields dealing with the deep past. Quine argued for the underdetermination of theories by *all* possible, available evidence and extended this argument as applicable to scientific knowledge in general. Although this may be an attractive option for certain philosophical discussions, it is immediately clear that addressing it can have little impact on the practical concerns of a scientific field and even may be an impediment for understanding the way the field operates.

In fact, a less abstract version of the concept of underdetermination predates Quine's work. At the beginning of the twentieth century, Pierre Duhem (1994, originally published in 1914) drew a conclusion about underdetermination of theories via concrete examples in physical sciences. When philosophical analysis of this sort is limited by details of concrete cases, it inevitably leads to more moderate conclusions. Duhem underscored the uncertainties inherent in observations, particularly in light of potential revisions to auxiliary hypotheses associated with any hypotheses under scrutiny. These revisions occur through observational and experimental tests, that is, through the evidence that is currently accessible. He argued that revisions, at least in theory, can be always outsourced to the auxiliary hypotheses, although this is often not pursued for pragmatic reasons (e.g., due to the adequate simplicity of a rival hypothesis) or because there is a temporary inability to make revisions with the available results.

Two major and opposed approaches to the underdetermination of rival theories by evidence in contemporary philosophy can be labeled logical and historical, based on their motivation and goals (Carrier, 2011; Pietsch, 2012). Although we find this distinction suitable for our purposes, others use the labels strong and weak (Newton-Smith, 1978), theoretical and practical (Butterfield, 2012, 2014), or Humean and scientific (Dawid, 2006). The logical approach ultimately aims at addressing the debate between the so-called realists and anti-realists on scientific knowledge. Very generally speaking, realists believe that posits of scientific theories denote real properties out there in nature, while anti-realists contend the business of scientific theories has to do with adequate predictions of events, and the correspondence of the sort realists assume is closed to scientific knowledge as a matter of principle.

We will not dwell on the debate predicated on the logical notion of underdetermination, as it would contribute little to what we think is epistemologically a much more interesting and methodologically more urgent question: the historical understanding of underdetermination. Some philosophers find the abstract, logical version of underdetermination is a more philosophically prudent pursuit, but as our analysis unfolds, we contend it will become clear that analyzing underdetermination in the historical context goes to the heart of our understanding of the scientific method, including the question of whether scientific theories grasp the truth about the natural world and if so, what that may involve. And it allows us to point out some very practical, epistemological consequences for day-to-day science. The logical approach is devoid of precisely defined rich, historical context. It is likely to simplify scientific pursuit, resulting in a simplified picture that, at the end of the day, may be far even from a plausible, idealized model of science and may have nothing to do with the actual scientific method.

Moreover, the attempts of the proponents of the logical approach to come up with algorithms for alternatives to existing theories are rightly rebutted by the critics of the underdetermination thesis, as they demand the presentation of viable alternatives. As Kyle Stanford (2006, 11) pointed out, it is reasonable to reject the thesis given the absence of convincing examples of well-confirmed competing hypotheses (models, or theories). The response of those who pursue the logical version of the thesis often turns into "argumentative efforts on the rather trivial forms of underdetermination" (Stanford, 2006, 11) and to the construction of overly ambitious algorithmic procedures that purport to generate alternatives for any given theory. First, we do not even know whether, generally speaking, scientific process can be subsumed under any general algorithmic procedures. And second, such procedures are bound to fall short in the "daunting task of generating alternative hypotheses that are both scientifically serious and genuinely distinct from competitors" (Stanford, 2006, 15), which the real scientific practice churns out.[21]

There is another substantial reason to go the route of historically based considerations of underdetermination. The logical approach is concerned with the assumption of all the possible observational evidence for a given theory. This idealized construct may be an exciting abstract philosophical proposition, but the problem is that substantial cases of underdetermination we encounter in real science rarely, if ever, fit an ideal case of complete empirical equivalence of the evidential domains underlying distinct rival theories, and this is presupposed in such highly abstract arguments. Rather, the evidential domains are equivalent to an extent, often in terms of key empirical results the theories purport to explain, but initially small disagreements and discrepancies in an often-wide domain of evidence eventually become decisive when judging one of the competing theories to be more adequate than another (Stanford, 2006; Pietsch, 2012). This key fact, along with the dynamics that underlies it, does not fit an idealized model of the logical thesis of the underdetermination of theories as predicated on total equivalence of their respective empirical domains. Such an assumption, although perhaps useful in some modeling tasks, is too coarse-grained to be applied to the studies of real cases. Thus, in real situations, certain key observations and phenomena will be interpreted differently by distinct theories as supporting different theoretical posits, while other key observational phenomena plausibly supporting one theory will be alternatively and also plausibly deemed an observational artifact of sorts in the other. The CMB case will reveal plenty of such situations, where overwhelming convergence on one theory took years or sometimes even decades.

Moreover, we cannot, as it were, come up with an arbitrary number of free parameters defining alternative theories that fit the existing observations – a situation fitting the idealized model – since these parameters are typically not quite independent. Free parameters are combined in coherent theoretical structures, and this severely constrains the number of real alternatives. Thus, in the case of the CMB's power spectrum:

Saying that the power spectrum contains hundreds of independent parameters for a given resolution is not correct, because different values of [angular power spectrum] are not independent in the same sense that hundreds of observations of the position and velocity of a planet do not indicate hundreds of independent parameters, the information on the orbit of planet being instead reduced to only six Keplerian parameters. (López-Corredoira, 2013, 1350032–4)

In fact, this idealized model of underdetermination misses the fruitful possibility of theorizing that practical underdetermination enables (e.g., computationally coming up with) *meaningful* alternative, free parameters fitting the observations. We will return to this possibility in Chapter 26.

Finally, we can never exclude the possibility that theories we discuss in a very abstract manner are actually only parts of a more general theory (Norton, 2008). Thus, we opt for an exploration of underdetermination deeply embedded in the historical context and its details, rather than being guided by particular philosophical assumptions that tend to construct the notion outside of relevant historical context.

Before we move to the analysis of the post-1965 cosmology, we should briefly explain how, generally speaking, the underdetermination situation has shaped cosmology in its historical context. Jeremy Butterfield (2012) addressed directly underdetermination in cosmology, making two general points. First, although cosmology as a field takes into account the uncertainties related to underdetermination, it deals with underdetermination of *models* of empirically well-founded fundamental theories, rather than with underdetermination of these theories themselves. Second, this underdetermination does not prevent a current realist acceptance of the orthodox model, namely the Hot Big Bang. Butterfield's argument represents one of the first substantial general discussions of underdetermination in cosmology among professional philosophers.[22] Our analysis will show, however, why his first point on the underdetermination of models as the sole focus in cosmology is only conditionally correct, and his second point on the realist embrace of the Hot Big Bang is correct but for all the wrong reasons. Unsurprisingly, both points are predicated on the usual textbook account of what happened after 1965 discovery of the CMB.

Regarding the first point, it is correct that debates in cosmology, including those on the origin of the CMB, are predicated on particular cosmological models stemming from fundamental theories. But a key feature of the quest for the origin of the CMB was a debate between various models predicated on opposing fundamental theories, the relativistic ones on the one hand and those espousing a Newtonian framework (broadly conceived), such as steady-state theories, on the other. For realists, the key question is when exactly the "thick evidence" designated by Butterfield as a precursor for the justified realist stand on orthodoxy is achieved, and more importantly, how thick is it. As Butterfield argued, if the evidence is thin we should be cautious (Butterfield, 2012, 7). He also very optimistically pronounced that a "great deal [of evidence for the Big Bang] is now established" (ibid.). As we will see, the latter pronouncement is correct, but was a judgment call and, as such, makes us rethink the idea that there is an identifiable universal red line dividing thin and thick evidence. Rather, what the study of the history of the CMB explanations tells us is that these thick lines are model dependent and context dependent, and different cosmologists working on different models (sometimes predicated on opposing fundamental theories) would have been entirely justified to espouse a realist attitude in the middle of these debates. Their acceptance slowly converged on a particular model, or rather a set of Hot Big Bang models, but we cannot identify a point where the realist attitude with respect to the orthodoxy became exclusively justified. In fact, as we will see, it should not have become so for a long time, given the context of the debate.

We will also show that Butterfield's claim that evidence of the Hot Big Bang is "today as certain as the fact that insulin has 52 amino acids" (Butterfield, 2012, 8)[23] is not a very helpful or informative view of an epistemological attitude prevailing in the day-to-day work of cosmologists. In fact, it may not be very informative in other fields of physical sciences. Even in the case of particle physics, for instance, "[t]o believe in the standard model in the early 1970s merely meant assuming that any more far-reaching change of physical postulates, in as much as it would be successful, would itself imply the standard model predictions" (Dawid, 2006, 7). This also applies to the current attitude of many string theorists (ibid.). Thus, such a claim is vacuous at best, merely stating that *certain key findings* will hold, even when, or if, the current theory is replaced. To be fair, Butterfield seemed aware of this and thus was cautious with his rather grand-sounding analogy; he clarified that the analogy applies to the "facts" we learn, and not to possible "conceptual change" or changes in the meaning of a particular theoretical concept. This leaves the door half open to future theory changes while seemingly cementing known facts that are supposed to justify the strong realist attitude toward a particular theory in a particular moment in the development of current cosmology but not toward the alternatives.

Yet herein lies the chief difficulty, and it prevents us from looking impartially at debates, like the debate on the CMB origin, and drawing solid epistemological conclusions. The distinction between *theory* and *facts* as used in this context is vague and often somewhat hastily used. As such, it is detached from practice and fails to square with the actual dynamics in the developments of science. What parts of scientific knowledge are contingencies and what parts are inevitable is an open question (Soler et al., 2016) and requires careful historical-philosophical case studies to be adequately addressed. This applies to post-1965 cosmology and to cosmology in general. In fact, a general argument on the distinction between theory and facts was made very early on by none other than Sir Hermann Bondi, one of the key cosmologists post-WWII. The argument is worth discussing in some detail as it provides some important distinctions and insights into the ways by which knowledge is generated in cosmology and points out the key misunderstandings to avoid in case studies.

Underdetermination has been a main concern of cosmologists, often very explicitly, although not labeled by this convenient term. This will become apparent in our discussions of the intricacies of the ties between various cosmological models and the CMB as the key evidence. Even so, it is useful to briefly introduce Bondi's early, deliberate, extremely refined, and intellectually dense effort to reflect on the epistemological foundations of the discipline in light of the omnipresent underdetermination.

Bondi made a comprehensive epistemological and methodological argument in 1955 targeting the widespread view of the speculative character of astrophysical theories, especially cosmological theories (Bondi, 1955). He argued, quite surprisingly, given the general understanding at the time, that inferences from observational research are not less prone to errors than theoretical inferences in cosmology. This, he said, was widely overlooked and had consequences for acceptance or rejection of the results (Bondi, 1955, 155). Moreover, he openly criticized distinctions between "observational fact" and "theoretical speculation" as misleading and detrimental to the field. Bondi also discussed the assessment of certainty across measuring techniques in labs: "In ordinary laboratory practice this assessment [of certainty] is generally based chiefly on past experience with the instruments concerned, combined with comparisons of their constructions and estimates of how 'direct' the measurements are" (Bondi, 1955, 155). Observations relevant to cosmology did not fair well in his analysis. First, he pointed out the theory-ladenness of seemingly simple observations illustrated using a photographic plate (ibid., 157). Second, he noted the problem of nonexistent repetition of observations due to lack of funding (ibid., 159). In a further effort to substantiate his view, he compared examples of theoretical errors of accepted theories, such as James Jeans's theory of instability of stars due to sensitivity to temperature (ibid., 158), to examples of observational errors, such as the observation of proper motions in extragalactic nebulae by Adriaan van Maanen and others that turned out to be incorrect by a factor of 100 or 1,000, and Sirius B shifts in spectral lines that are impossible to measure due to obscure pressure shifts (ibid., 159).

The motivation for Bondi's meticulous argument on the uncertainty of observations should not be surprising. After all, he was one of the experienced physicists working on cosmological problems, along with Hoyle, who also knew full well that errors in observations are not always necessarily random, but often prone to systematic expectations that creep into the observational process. Hoyle (1993, 178) described a poignant moment when he connected this possibility to the early measurements of radar. He realized that "the measure of hope" of the working scientists resulted in the nonrandom error he discovered, and others should be very wary of this. Then, later on, when Bondi was writing his paper, and they were communicating on the issue, Hoyle sifted through publications on cosmological topics and concluded "observational mistakes were far more common than theoretical mistakes, suggesting that the refereeing process was less good at weeding out poor observations than it was in detecting poor theory" (Gregory, 2005, 43).

Interestingly, Bondi made a general argument in which he drew a distinction based on his comprehensive analysis and examples. Thus, there are two aspects of work in cosmology: *observational inferences*, immediately and directly drawn from novel data, and *general theoretical inferences* that use terrestrial physics to explain astronomical phenomena and resolve astronomical problems (Bondi, 1955, 155–156). Observational inferences use terrestrial physics to a certain point, in as much as theoretical inferences inevitably rely on certain observations, but they are drawn at different points of the theory-observation interface. Bondi quite explicitly acknowledged a much more substantial underdetermination of observational inferences than of theoretical ones, stating: "In observational work long chains of inferences are based on frequently somewhat uncertain data, whereas in physical theories of astronomy, though long chains of inferences are also used, they are generally based on much more reliable experimental data" (ibid., 158). To this he added: "The purely factual part of the vast majority of observational papers is small. It is also important to realise that these basic facts are frequently obtained at the very limit of the power of the instruments used, and hence are of considerable uncertainty" (ibid.). Thus, if observational facts in cosmology are typically at a limit of observational powers, this will render theory predicated on them too provisional to draw firm red lines of before-and-after periods of unjustified and justified realist attitudes.

Bondi continued by challenging the parlance that pervades epistemological descriptions of scientific knowledge in both science and philosophy. Dividing the work into observational facts and theory is not only tendentious but can also be harmful; it often leads to hasty and poorly justified conclusions:

To refer to observational results as "facts" is an insult to the labours of the observer, a mistaken attempt to discredit theorists, a disservice to astronomy in general and exhibits a complete lack of critical sense. Indeed, I would go so far as to say that this sort of irresponsible misuse of terminology is the curse of modern astronomy. Present-day astronomy may fairly be called the science of extracting the maximum information from the fundamentally meagre data that can be obtained about outer space, an endeavour to stretch both observation and interpretation to the very limit. (Bondi, 1955, 158)

This distinction may be acceptable as a rough approximation but it is an obstacle to a deeper understanding of cosmology.

If Bondi's points are sound given the context of research in cosmology, then Butterfield's notion of stubborn facts and ensuing theory does not bode well for the study of actual history. But how did this epistemological situation change in the decades after Bondi made his argument? The crucial pieces of observations steadily changed over time, leading to dynamic theoretical debates. Thus, for instance, as we will see, the key measurements of the shape of the CMB spectrum, that we would be hard pressed to label as facts in Butterfield's strong sense as it is conveyed by his analogy, went back and forth in terms of the particular frequencies deviating from the blackbody spectrum shape. These twists in the observational domain were a crucial factor in the generation of new versions of alternative explanations of the CMB origin. Thus, generally speaking, a crude fact/theory distinction is simply not helpful in understanding the dynamics of the Hot Big Bang as the winner, even though it may provide justification of a very particular realist attitude toward the Hot Big Bang at a particular point in time (certainly after the *COBE* measurements and even more so today).

Bondi's distinction between observational and theoretical inferences is also a useful way to divide the unorthodox explanations of the CMB and understand their respective motivations. As we will see, some explanations were based on the latest observational results and did not attempt to build a full-scale model, while others derived predictions from already built models or accommodated the latest results. His early criticism informed many subsequent debates and criticisms of our cosmological knowledge and approaches, from those still subsumed within the great controversy to those of Disney (2000), López-Corredoira (2014), or contemporary critics of inflationary paradigm like Ijjas, Steinhardt, and Loeb (2014).

To sum up, the extent of underdetermination inevitably changes over time in any scientific field. Underdetermination is almost always inherently transient, but in fields such as cosmology, especially in their early phases of development, the periods of underdetermination last long enough to shape methodological ramifications, the nature of inferences, and the nature of the debates. It takes a long time to provide reliable detecting apparata and techniques and then to adequately digest the incoming data within the network of existing reliable theories and new theoretical ideas (many of which are necessarily half-baked or improperly formulated, or perhaps just too exotic at first). This is why, as we will show, the convergence of observations and theory to the Hot Big Bang was much slower than usually thought. Any shortcut would have suffered from the epistemic deficiencies that Bondi's analysis pointed out, and the arrival at the stubborn facts phase Butterfield bet on would not have been possible. We draw some general conclusions on underdetermination in cosmology in Chapter 24, after we discuss the crucial details of the alternatives to the CMB orthodoxy (including their interaction and construction), how underdetermination played out in practice, and how cosmologists and physicists reflected on such a state of affairs.

9

Was the CMB a smoking gun?

Before we move on to the study of the CMB explanations in Chapter 10, there is another assumption that can impede analysis if not properly addressed. It is the notion that the CMB was nothing but 'a smoking gun' that scientists were looking for all along. This is another assumption whose consideration goes to the heart of the methodological situation in cosmology.

As we have stated, the discovery of the CMB belongs to a family of, broadly speaking, deep historical scientific fields dealing with unique phenomena occurring in the deep past. In this respect, it is similar to discoveries such as the iridium anomaly at the Cretaceous–Paleogene boundary by Luis Alvarez and coworkers in 1980 or the Ediacaran biota by Martin Glaessner in 1959. The phenomena these sciences deal with are typically not prone to the desired level of experimental manipulation, if at all. To this methodological limitation, we can add the unique physical limitations of cosmology that Ellis (2014) emphasized: the speed limit, the observational horizon (about 42 billion light years distant), the need to test physical laws in the past, and various selection and detection effects (e.g., failing to detect sources, or objects that do not emit significant radiation or do not radiate at all).

Generally speaking, because of the field's aforementioned traits, the ground for multiple interpretations over the decades of the inception of modern physical cosmology was richer than it was, for instance, in various areas of experimental physics. Thus, it should not be surprising that the domain of the alternative interpretations of the CMB was quite wide. As we pointed out earlier, the CMB was a milestone discovery. But the way it came about and the way it was assimilated into the standard interpretation was not analogous to, say, the evidence that a particle collider provides to break the stalemate between competing theoretical accounts in particle physics. The convergence process unfolded much more gradually and tenuously.

More importantly, the epistemological standing of both orthodoxy and the alternatives can be portrayed as analogous to a field such as experimentally driven particle physics only in a fairly simplified manner. The sidelined alternatives in a scientific field like cosmology, with all the characteristics listed above, may eventually have the same standing as alternatives falsified by repeated controlled precision experiments, but what happens on the way toward that ideal goal is crucial for understanding how that field actually works. The process of challenging the consensus and coming up with multiple sticking points that challenge it and ought to be resolved is protracted, preventing a desired definitive judgment.

Carol Cleland made a series of arguments (2002, 2011) in favor of the view that historical sciences dealing with the origins of phenomena through indirect evidence, or rather, traces of phenomena they theorize about, are typically in the business of seeking a smoking gun for the models they propose. She summarized the methodological approach of historical sciences as focusing "attention on formulating mutually exclusive hypotheses and hunting for evidentiary traces to discriminate them. The goal is to discover a smoking gun" (Cleland, 2002, 480). Indeed, it is enticing to interpret the discovery of the CMB as the smoking gun – or to push the analogy for more precision, the smoke released from the gun – that the proponents of the Big Bang were seeking all along and that was finally found in 1965.

Moreover, in terms of scientific objectivity, seeking for traces, that is, of smoking guns, is on a par with experimental work (Cleland, 2002, 476). Such a search is analogous to a protracted series of experiments versus a single experiment in physics that suffices for a test of a particular hypothesis (like the experiments of Michelson-Morley, Eötvös, or Aspect et al.). The difference lies in the extent of elaboration, not a substantial epistemic quality. Cleland's initial admirable motivation for making such a strong claim was to counter a litany of rather crude and often derogatory methodological critiques of historical hypotheses in natural sciences. She cited the following particularly egregious example from an editor of *Nature* who said, "They [historical hypotheses] can never be tested by experiment, and so they are unscientific" (Gee, 1999, 5–8).

The smoking gun view of the methodological approach in historical sciences works to an extent in cosmological debates once we zoom out of the details of the actual debates. Once we pay attention to the dynamics of the debates and how exactly the proposed models were assessed within the community, it will become apparent that the creators of the models and explanations often endorsed the smoking gun view, albeit implicitly or reticently. Indeed, they were often searching for one. What will become apparent as well, however, is that discovery of the CMB did not clinch "the case for a particular causal story" (Cleland, 2002, 482), as Cleland thought a smoking gun did, but it certainly resulted in a number of serious challenges to it. Even some immediate proponents of the Hot Big Bang model worked on alternatives, testing its weaknesses, obviously not ready to pronounce the newly found radiation as definitive evidence of the Hot Big Bang.

Thus, a sole focus on the smoking gun approach would prevent us from fully understanding the debate and its intricacies. (Perhaps the very *smoke* from the smoking gun prevents us from seeing who actually pulled the trigger?) What starts as a simple model or an explanation, perhaps in search of a smoking gun, often grows to a complex one, quickly exceeding the powers of such a simple strategy of seeking evidence (Currie, 2013). Or scientists may not start with a unifying model at all but attempt to explain a particular phenomenon with a set of other well-known phenomena (Currie, 2013). We will discuss an illustrative example of the former, the Hot Big Bang model as it developed over decades, and Narayan C. Rana's tired starlight explanation of the CMB origin as a telling case of the latter. So yes, the CMB may have been perceived as a smoking gun by *some* of the early creators and proponents of the Hot Big Bang model, but this epistemic attitude was quickly swept aside by the burgeoning debate and the development of the orthodox model itself.

Moreover, and more importantly, the distinction between methodological approaches in experimental and historical sciences seems to be a matter of degree – who uses which method and to what extent – not a clear dividing line. Although she may have thought experiments and trace-reconstruction were equally potent epistemically, Cleland would disagree that methodologically they are distinct only as a matter of degree. She contrasted the two "as evidential reasoning … from causes (test conditions) to effects [experiments], with the concomitant worries about ruling out false positives and false negatives, and from effects (traces) to causes [trace-reconstructions], with the concomitant worries about ruling out alternatives" (Cleland, 2002, 484). Again, there is an aspect of cosmological searches that can be identified as the latter, primarily as an attempt to eliminate alternatives. Yet once a particular hypothesis or a model is established, scientists turn to an approach that is typically understood as experimentalist, including Cleland's characterization of it.

As we will see, a series of observations of the CMB over the years were keenly sifted through for indications of false positives in the existing evidence to test for possible substantial variations in the shape of the spectrum, as this would remove it from the shape of the blackbody radiation predicted by the orthodox model. And each new batch of evidence for a blackbody shape of the CMB spectrum was followed by further observations in changing conditions (e.g., moving from ground observations to observations from balloons and rockets and observations from satellites). Moreover, such observations had all the marks of controlled experiments; for example, flying a jet fighter equipped with spectroscopes in different directions to eliminate the potential artifact signals produced by the immediate background.

Any discussion of the nature of experimental and observational inferences is bound to be a complex and important debate in its own right, so for the purposes of our analysis, it suffices to say that scientists working on the explanations of the CMB origin alternated between historically and experimentally minded modes of reasoning, as Cleland outlined them. In fact, if we take a closer look at modern experimentation in particle physics performed in giant colliders, these two modes of reasoning are more or less continuous and often difficult, if not impossible, to distinguish. The physicists in accelerators deal literally with the traces of events they postulate. They analyze marks (or the lack of marks in the case of noninteracting neutrinos) of what they think are particular particles that leave very specific traces in detecting equipment (film emulsion or electronic circuits) and infer the properties of these indirectly observed particles by comparing them with their hypothesis-driven simulated events.

The most informative difference between a fully experimental approach and historical reconstruction in cosmology, however, is not related to the purported sharp divide between these two modes of reasoning, if it exists at all, instead of a continuum or constant alternation of the two. Either way, distinct, continuous, or alternating, they are in the business we call science. Rather, the key difference between heavily experimental scientific fields and scientific fields like cosmology is *the extent* to which scientists in their respective disciplines studying their phenomena of interest can test, that is, *control for* and *manipulate* desired parameters. Perhaps this is the sole substantial difference (Perović, 2021).

The potential for testing desired parameters is vastly superior in what we normally call experimental sciences – such as experimental particle physics or inorganic chemistry – than it is in cosmology. Since the early experimental work on the physics of elementary particles, physicists have aimed at determining the basic parameters, starting with mass, electrical charge, and spin, later turning to other charges of both (relatively) free and interacting particles. They have been able to successfully vary the angles of particle interactions and the density of their streams at extremely different levels of energies. Various ways of producing these streams, as well as their collisions with fixed targets or with other streams, and very different detecting techniques have been devised from the turn of the twentieth century until today. Natural phenomena have been manipulated and intervened in to an unprecedented extent, and the parameters of our theoretical accounts have been similarly varied as we seek to understand them better. (Experiments in chemistry and biology, as manipulative as they may be, take place at roughly the same constant level of energy unlike high energy physics experiments.)

Now, imagine if particle physics were confined to unique events that physicists could not produce serially. In other words, imagine if physicists were confined to mostly observing phenomena as they develop in nature without the possibility of interventions and manipulations. This sort of knowledge would be continuous with the knowledge in actual particle physics, and physicists could use observational (in astroparticle physics) or computational surrogates of fully experimental techniques, but it seems clear that the challenges would be very different, and the key insights would appear in a much slower fashion, if at all. The gulf between such knowledge and the knowledge we have actually obtained would be vast.

Conversely, the analog in cosmology of actual experimentation in particle physics would be the possibility of varying each cosmological parameter that we may think defines the CMB properties (its temperature, spectrum shape, isotropy, polarization…). Yet we cannot vary the rate of the expansion of the universe to see how it affects the temperature of the CMB, nor can we vary the density of intergalactic dust to see how it affects the thermalization of cosmic background radiation. We can only observe relevant phenomena in clever ways, in ways that methodologically resemble experimental reasoning, mostly using instruments located below the atmosphere or a few telescopes outside it (this difference was crucial in the string of observations on the CMB and the emergence of unorthodoxy).

It is indeed possible (Boyd & Matthiessen, 2023) that epistemically, in principle, the experimental and observational approaches to understanding physical phenomena are comparable, and in certain contexts, observational data can even offer advantages. The commonly highlighted inherent advantages of manipulating phenomena in experiments (Okasha, 2001; Zwier, 2013; Currie & Levy, 2019) may, in some cases or even in principle, provide only practical benefits, while observational techniques could compensate in other aspects. For instance, in the realm of astrophysics, there is a multitude of available objects for observational study, often sufficient to compensate for the absence of direct physical manipulation. However, this advantage becomes only tangentially relevant in the context of cosmological knowledge, as the unique nature of the subject of study prevents circumvention of its inherent physical limitations.

Furthermore, as our analysis will reveal, this intrinsic limitation interacts with potential confounding factors (explored by alternative accounts of the CMB) within the realm of cosmological phenomena, resulting in a particularly complex landscape for cosmological study. Thus, briefly stated, experimental physics and cosmology are very close to being at the opposite ends of a continuum extending from limited and passive observations to full-scale intervening experimentation, with a concomitant ability to access natural phenomena to the desired extent (Perović, 2021). And this constitutes the main difference between mostly historical and experimental sciences. In the case of the former, a comprehensive and decisive testing of the central parameters of models is typically difficult or impossible – the Big Bang has likely occurred only once after all – and typically wanting, thus making the prolonged state of underdetermination in cosmology a regular affair. Historical science is confined to more transient underdetermination episodes than experimental science, and this, as we will see, has crucial epistemological and methodological consequences.

We should note that despite the difference in the longevity of underdetermination due to the primary available method of investigation in each field, because financially and technically very demanding experiments are lagging decades behind theoretical developments, the current situation in physics may be much closer to that in cosmology than any time previously. It took about three decades to test the Higgs boson hypothesis, a length of time unimaginable in the particle physics of the late nineteenth and early twentieth centuries. The two disciplines can be even meaningfully compared at this point. As Richard Dawid pointed out:

Particle physics today deals with a number of speculations such as grand unified theories, dynamical electroweak symmetry breaking or large extra dimensions. Cosmology also comprises a wide range of speculative models. In all these cases, scientists develop educated guesses about the likelihood of the concept's future success by assessing the amount of indirect empirical support, available alternatives, the concept's inner coherence and potential power, its simplicity, aesthetic attractiveness, and so on. (Dawid, 2006, 3)

That particular and rather unusual situation in physics aside, it may be epistemically imprudent to treat alternative interpretations of the CMB that have been sidelined for observational and theoretical reasons the same way we would (justifiably) treat alternatives falsified via an elaborate experimentation process in physics. Rather, we should regard them as a resource that can potentially be revised and revived despite occasional straightforward falsifications of certain aspects. Thus, the fact that evidence for the orthodoxy in cosmology is evidence in a field dealing with deep natural history should on its own stop us from the wholesale discarding of the alternatives.

Moreover, reinforcement of the orthodoxy has occurred in cosmology by propagating an inadequate history of the field in concert with the systematic, unjustifiable downplaying of many alternatives that, in fact, provided a necessary springboard for the development of the orthodoxy in the first place. Generally speaking, failing to understand the crucial details about the history of the orthodox view of the CMB leads to the widespread misperception that the dissenting views have been both few in number and insignificant in content.

10

Classifying and analyzing unorthodoxies

Before we begin discussing the actual alternatives in more detail, we note that brief historical discussions of the CMB interpretations can be found in the literature (e.g., Norton, 2017). In addition, before discussing their own unorthodox interpretation of the CMB data available at the time, David Layzer and Hively (1973) divided the known interpretations into two groups along lines similar to those we employ. Yet systematic, historical, technically informed comparative analyses are lacking, especially those drawing important epistemological and methodological conclusions, and in light of the current *WMAP/Planck* data, those keeping an eye on future developments.

We suggest a taxonomy of alternative explanation of the CMB to help the reader follow the argument and to emphasize their defining features (see Figure 10.1). First, they can be divided in two broad categories: (1) those predicated on the acceptance of the cosmological validity of field equations of general relativity, and (2) those abandoning it in favor of a more or less different conceptualization of gravitation and cosmological spacetime.[24] The former are constructed within the framework of Friedmann models, explicitly or implicitly (see Appendix A). Some stick to this framework only as a first approximation limiting a toy-model interpretation. Prominent examples belonging to this group that we will discuss in some detail are the Cold and Tepid Big Bang models (e.g., Carr, 1977; Carr & Rees, 1977; Aguirre, 1999). The interpretations belonging to this category veer far from the orthodox view in many respects, but they are relatively moderate alternatives when compared to the second category, as they share some key features with the standard interpretation (e.g., expansion of spacetime from a singular or near-singular state, large-scale homogeneity and isotropy, our typical position as observers, growth of gravitational perturbations as the origin of cosmological structures, and many others). In contrast, radical alternative interpretations challenge even basic cosmological assumptions of the standard view. This is especially true of those interpretations of the CMB within purportedly nonrelativistic world

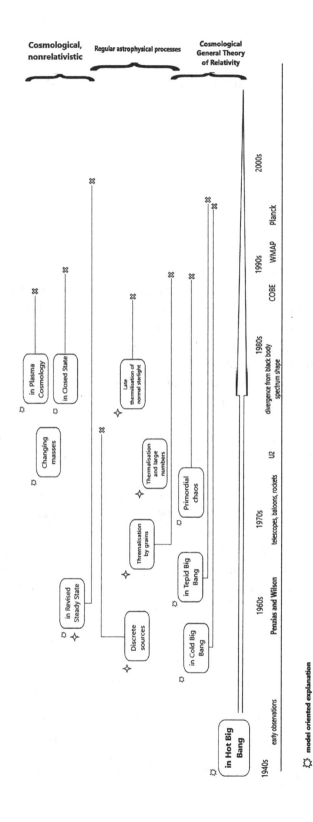

Figure 10.1 Various explanations of the origin of the cosmic microwave background from 1965 onward.

models, including various steady-state cosmologies that *apparently* violate even the conservation laws of laboratory physics (Bondi, 1960, 146).

Our suggested taxonomy reflects distinct astrophysical and cosmological motivations for constructing the accounts. This is in accord with the general tendency to approach problems from two different angles in modern cosmology: astrophysics and particle physics respectively. This divide has acquired distinct methodological and sociological aspects, some of which were actually commented upon by the participants in the CMB debates, notably Hoyle (1994, 256–305).

In our analysis, we will look at: 1) the technical details of models and explanations; 2) the predictions and tests they almost invariably suggested, or accommodations of the latest results[25]; 3) the exact historical context, including the state of relevant knowledge at the time, as well as the authors' professional and institutional backgrounds; and 4) explicit and reticent epistemic, metaphysical, and methodological motivations. In 4), we will discuss the main philosophical aspects of the alternative models and theories, focusing on the explicit epistemological attitudes and epistemic ramifications of constructed alternatives, as well as on more implicit and reticent ones that motivated the research. The analysis will make it apparent that the alternatives greatly differed in terms of the style and kinds of arguments they developed, from very abstract theoretical constructions to constructions developing relevant physical details, and to those modifying existing elaborate models. Finally, we will identify the questions they posed, both those directly challenging the emerging orthodoxy, and those that were placed on a back burner, some of which have remained relevant (in a modified context) until today. Even more, some of the questions posed by the alternative CMB models have, as we shall see, already demonstrated capacity to become relevant after a historical hiatus. Thus, it is quite possible that some of these questions may resurface again in the future.

Part IV

Moderate unorthodoxies: The CMB with the Big Bang

11

Cold and Tepid Big Bangs

Population III objects

The prime examples of moderate unorthodoxies are various versions of the Cold and Tepid Big Bang cosmological models. They share the assumption of the singular origin of the universe in relativistic Friedmann models with the standard Hot Big Bang models, but they introduce very different initial conditions, among which the low or intermediate value of the photon-to-baryon (or entropy per baryon) ratio η stands out (see Appendix A). As this chapter explains, these models had certain advantages over the Hot (high value of $\eta \sim 10^9$) Big Bang models, especially with respect to the initial conditions and their implications.

Models with low photon-to-baryon ratios developed over the last five decades are commonly labeled Cold or Tepid Big Bangs (Zel'dovich, 1963, 1972; Layzer, 1968, 1992; Barrow, 1978; Aguirre, 2000).[26] An initial advantage of these models is their more plausible account of structure formation than the one in the Hot Big Bang model; whichever way the initial perturbations are generated in the models, they are intuitively – and formally – easier to grow into the galaxies we see today and into the current overall hierarchical large-scale structure. A lower temperature (hence lower pressure) at the same matter density is more conducive to the growth of gravitational perturbations than a higher temperature. Cold and tepid models also avoid the overproduction of monopoles and other topological defects via the Kibble mechanism (cf., Kolb & Turner, 1990) and resolve several problems related to the phase transitions in the early universe that orthodoxy has to deal with. The question driving this alternative was explicitly posed by Charles M. Misner (1968, 433): How do homogenous initial conditions produce small irregularities that result in galaxies? He addressed the question with a discussion of the origin of anisotropies in his paper.

In a somewhat more radical interpretation of the CMB, the models of bouncing universes treat the Big Bang not as an initial singularity, but as a bounce from the previous contracting epoch. Although largely forgotten today, these models were a popular alternative in the 1970s and 1980s.[27] They are also an example of the return or reusage of abandoned alternative ideas. They ran into difficulties with respect to the entropy production in the epoch preceding the bounce, and they lacked a convincing mechanism for the bounce itself, but similar models are now proposed in string and loop-quantum cosmologies based on certain approaches to quantum gravity (e.g., Millano, Jusufi, & Leon, 2023). There are several instances of the reusage – or *exaptation*, in the parlance of evolutionary biology[28] – of intriguing ideas that first appeared in discussions about the CMB origin. Such ideas may be reused on a purely rhetorical level to artificially extend the lineage of an idea farther into the past, similar to the motivation for many public actors in the eighteenth and nineteenth centuries to seek precedents in antiquity.

There is, however, a more important cognitive and methodological reusage at play. This practice is seldom studied within the framework of history of science, and modern physical cosmology offers us one of the very best documented case studies. We return to this topic in the concluding chapters.

Our aim is not to defend or resurrect any clearly refuted cosmological model, but it is useful to distinguish the models' theoretical features that directly address the origin of the CMB. We can analyze that aspect independently without discussing and weighing the strengths and weaknesses of the models' other theoretical features. In fact, the need to identify a physical mechanism separating the CMB origin from the Big Bang is their central feature: if we assume the premise of a Cold or Tepid Big Bang, it is necessary to identify a source for more than 99% of the photons that currently exist in the universe, as well as a source distinct from the Big Bang itself. Such sources are hypothesized as Population III stars.

We should note right away that the technical astrophysical term "Population III" refers to the first generation of stars without any metal content: $Z = 0$ (Carr, 1994). These primordial stars were composed of the output of primordial nucleosynthesis, that is, hydrogen (both H and D) and helium (mostly ^{4}He, but with a small fraction of ^{3}He) and negligible amounts of ^{7}Li. According to the standard view, these stars started the chemical evolution of baryonic matter. Yet when considering the cosmological models fashioned after 1965, it is prudent to include any objects that may have formed in the early epochs and became (quasi)stable early on. Therefore, when we refer to Population III objects, we assume this wider meaning to include, aside from the Main Sequence stars, pregalactic stars and black holes or even magneto-turbulent pre-galactic gas clouds (Layzer, 1968).

There is a need for conceptual precision here. It is one thing to conclude that if all chemical elements heavier than lithium (astrophysical "metals") originated in stars, and if all baryons were in the form of cooling and expanding plasma at one point in the early universe, there logically had to be a first generation of stars without any metals, that is, Population III. It is another thing entirely to set particular properties of those objects – which have not yet been observed – within an explanatory hypothesis or scheme. Making a clear distinction between the two would excise some of the confusion reigning in this domain and perhaps demonstrate the relevance of careful philosophical analysis.

The domain of Population III objects is not simply a matter of convention because the physical reference of the notion is somewhat unclear. The size of the first objects to undergo gravitational collapse remains highly disputed. The Jeans mass is typically calculated for each epoch of the early universe, and there is some controversy over which parameters should enter the calculation. The exact role of the Jeans mass in the formation of compact objects is also controversial (e.g., Gnedin, 2000; Schneider et al., 2003). Thus, the characterizations of Population III objects are inevitably provisional, driven by the goals of the study. They can hardly be characterized as part of the core knowledge in the standard model. In any case, the Population III objects are necessary components of the standard cosmological picture, as some metal content had to be formed first in such objects; the exact properties that should be detected are still vigorously debated as active research programs unfold.

A brief survey of theoretical, methodological, and wider epistemological attitudes that drove these accounts of the CMB, as well as attempts to combine their features, reveals their surprising diversity and ingenuity. In 1968, Misner paraphrased William A. Fowler when he commented:

The cosmological interpretation of this 3° K is not so much established as it is unchallenged, and an alternative explanation is not ruled out on energetic grounds since recent burning of hydrogen to helium in stars could supply the necessary energy if mechanism could be found for thermalizing it. (Misner, 1968, 432)

Following this motivating line of argument, Misner suggested the "prediction" of the current universe, that is, producing a hypothesis by reverse-engineering a process, with alternative relativistic theories, offered a recipe for constructing one, and he drew a general distinction between two possible strategies. One strategy might try to "show that almost all solutions of the Einstein equations which lead to star formation also have many other properties compatible (or incompatible) with observation" (Misner, 1968, 432). A more modest strategy would "attempt to survey much more limited classes of solutions of the Einstein equations to see whether some presently observable properties of the universe may be largely independent of the initial conditions admitted for study" (Misner, 1968, 432).

More than a decade later, this somewhat skeptical attitude toward orthodox interpretation was still a motivating factor for building moderate alternatives. In an explicit expression of the epistemic attitude that motivated his model, Hannes O. G. Alfvén stated that in contrast to the vigorous criticism of the static universe models, "no decisive objections against [the Big Bang hypothesis] were presented forcefully enough" (Alfvén, 1979, 23). He developed this essentially socio-epistemological analysis, obviously aiming to make a methodological point, by adding that because they were free from substantial criticism, the proponents of the hypothesis developed it in various directions, instead of bothering to question its key assumptions. As a result of this laissez-fair approach, "the large body of observations which are not in agreement with the Big Bang hypothesis are either neglected or accounted for by a large number of *ad hoc* hypotheses" (Alfvén, 1979, 23). In particular, Alfvén pointed out that the Big Bang model does not follow from Friedmann models, since no pressure gradient is defined in it; rather, it is added as an *ad hoc* postulation (Alfvén, 1979, 27). Resorting to his socio-epistemological complaints about the model, he argued it is "associated with ex nihilo creation of the universe," and "[a] discussion of a possible 'ultimate cause' belongs to philosophy or religion" (Alfvén, 1979, 32), not to physical cosmology.

In their work, Bernard J. Carr and Sir Martin Rees (1977) focused on two key assumptions of the Hot Big Bang model. First, they emphasized the scenario of present-day universe inhomogeneity as generated from small fluctuations as applicable to the Tepid Big Bang only. Second, they pointed out the implausibility of cosmological entropy in the Hot Big Bang – that is, the density of matter and radiation are comparable only by coincidence during thermalization, in contrast to the Tepid Big Bang scenario. Furthermore, the present photon to baryon ratio could be observed if the universe started with a much lower ratio (i.e., low entropy, due to tepid early universe), as it would evolve to the present ratio, and the accretion of massive black holes originating in it would provide the bulk of the CMB. Early condensation and formation of structures would occur because of low temperatures – massive black holes were bound to form early (before 10^{13} s), and as a result, radiation from the accretion of surrounding gas thermalized as the CMB. Carr and Rees (1977) also offered a detailed account of the dependence of massive black hole formation on the photon to baryon ratio (see Figure 2 in their paper).

In yet another general criticism pointing to a key epistemological deficiency of the Hot Big Bang model, Alfvén and Mendis (1977) made an explicit case for the underdetermination of cosmological models in light of the existing evidence, by pointing out the impossibility of observing the metagalaxy at earlier than $Z > 40$:

We stress that the observed cosmic microwave background radiation does not give us direct information about the state of the metagalaxy further back than the epoch corresponding to $Z \approx 40$, when galaxies or protogalaxies already existed, and when its own temperature was only about 110 K. To claim that it is strong evidence in support of the Hot Big Bang cosmology whose earliest epochs correspond to $Z \gtrsim 1,012$ with corresponding temperatures of matter and radiation $> 1,010$ K, is completely unjustified. (Alfvén & Mendis, 1977, 699)

In his 1979 paper Alfvén (1979, 24) argued it is plausible that the metagalaxy had an inhomogeneous density and was also hierarchical (i.e., composed of sub-clusters (Alfvén, 1979, 30–31)). Noting possible intrinsic variations of the Hubble constant, he suggested that in an extended Big Bang, rather than expansion from an initial singularity, redshifts will appear eventually when our protogalaxy leaves a sphere where initial explosions of other protogalaxies have occurred.

The criteria of nucleosynthesis and the temperature of the early universe distinguish the Cold Big Bang models from the merely Tepid ones (see Appendix A). These alternative Big Bang models aim at explaining the CMB without a dense, hot phase near the beginning. Carr (1981b) explained the motivation for such proposals in the following way:

In proposing that the 3 K background is generated by pregalactic stars or their remnants, we are not necessarily assuming that there are no photons before the stars form. This would be unrealistic since many processes in the early Universe would inevitably produce some radiation (e.g., the dissipation of initial density fluctuations or primordial anisotropy). Furthermore, a Universe which was "cold" at the neutron-proton freeze-out time $(\sim 1 \text{ s})$ would be unlikely to produce the observed 25 per cent helium abundance through cosmological nucleosynthesis (although the helium might, in principle, be synthesized in pregalactic stars). (Carr, 1981b, 671)

In an earlier paper Carr (1977) argued Hot Big Bang models predict an excessive formation of black holes, but this is inconsistent with existing observations. In contrast, a cold photon-less early universe would produce them consistently with the observations, while black hole accretion would provide the source of heating of the universe after the production of hadrons. He was honest about an apparent disadvantage of his model – its implausible assumption that primordial stars produce the observed abundance of helium was a prediction that could eventually falsify it. This is another instance of the exploratory nature of much of the Population III research in the 1970s and 1980s, when on the one hand, very little empirical data on the high-redshift universe were available, and on the other hand, nonstandard, alternative cosmologies were much livelier than after the solidification of the orthodox view.

Along these general lines, Carr (1981a) developed a very detailed account of various plausible combinations of relevant parameters of early black hole accretions and plausible phases of these evolutionary trajectories. He explicated underlying assumptions he argued were plausible for both black hole accretion and cosmological details. The values of basic parameters of temperature, density, and location of black holes in the galactic disk result in five distinct phases: 1) in an early phase dominated by radiation, there is little accretion; 2) in a pre-recombination phase, the accretion of matter is faster than that of radiation; 3) in the "Eddington phase" (the Eddington limit is the maximum luminosity of a body in a state of balance between the inward gravitational force and outward directed radiation), large black holes are accreting matter; 4) in the "Bondi phase" just before the formation of galaxies, the heating effect of black holes decreases (following Bondi's formula for accretion); and 5) in the final phase, the mass of the universe plausibly forms cold "blobs." He then considered various scenarios of the resulting 3K background radiation following each phase. Not surprisingly, a follow-up to this paper (Carr, 1981b) was prompted by the measurement results of Woody and Richards (1979) showing a deviation of the CMB from the blackbody spectrum shape, leading to the optimistic expectation of a Tepid scenario that was then regarded as a possible natural setting for grand unified theories (GUTs) of strong, weak, and electromagnetic interactions (Carr, 1981b, 671).

Alfvén (1979, 25) stated that the claim of homogeneous structure figuring in the Big Bang theory could not be proven or disproven within the realm of the evidence available at the time – the claim is true if quasars are at cosmological distances and if variations in galaxies' isotropy are explained by a *d hoc* auxiliary assumptions. He argued some quasars may not be at a cosmological distance, making the work of Burbidge, as well as emerging radio and X-ray data, relevant (Alfvén, 1979, 32–34). We return to Alfvén's views on plasma cosmology in Chapter 22.

In 1973, David Layzer and Ray Hively developed a Cold Big Bang model based on a detailed account of the CMB's origin. This model introduces early Population III stars as the ultimate source of energy of the CMB photons. The motivation for this theory was somewhat different from that underlying Layzer's previous theory (1968). The theory is discussed more fully in Chapter 13. For present purposes, we simply mention that a key sticking point of their alternative account (and multiple other unorthodoxies) concerned the abundance of helium in the early universe. How much of it was burnt early on? A lot was needed for early vast radiation for the thermalizing grain hypothesis to obtain. Layzer (1968, 101) had already discussed how much He was present in the stars and at what exact point. If shining stars comprised a much bigger portion of the universe in early epochs than they do now, hydrogen fusion in stars would suffice in terms of the actual energy produced. If the CMB was indeed powered by such stars, then dark matter must consist mainly of stellar remnants (e.g., old neutron stars and non-accreting black holes) of those early populations and not only of baryonic matter.

However, this was not easy to test even in a preliminary fashion, as the theory of primordial nucleosynthesis was not well developed at the time. Authors could not develop specific predictions of maximum baryon abundance that would be consistent with the observed abundance of light nuclei. As a result, they grossly overshot the value of Ω_b. Yet the key motivating point driving their accounts was that early fluctuations that resulted in the formation of galaxies had a natural solution in the

Cold Big Bang: internal density energy is negative while cosmic pressure vanishes under constant temperature (Layzer, 1968, 101–102, 1992).

It is not surprising that the Cold Big Bang models have not been as vigorously criticized or faced as much hostility as the steady-state models and later variations. Although this may seem to reflect sympathy for the Cold models, it also means their potential implications and weaknesses have not been studied as thoroughly as those of the Hot models. Rather, they have been perceived as "toy models," crafted to probe and ultimately emphasize the advantageous features of the standard model. In fact, this view was reinforced by Yakov B. Zel'dovich (1972, 1980) and Rees and Carr (1977) who initially promoted the models but then switched fairly quickly to the standard paradigm.

Alfvén's alternative model (1979) of Hubble expansion in a Euclidian framework is a good example of a toy model strategy. Alfvén aimed to develop an alternative as far as possible from the standard assumptions of Hot Big Bang but in accord with the existing observational data and without any new laws of physics (Alfvén, 1979, 23). This was a transcendental argument, in the parlance of philosophy, as he was aiming to identify necessary conditions for a particular state of affairs and to establish an exploratory model: instead of a succumbing to a "strong drive" to accommodate data to the standard Big Bang, he said, "it seems justified to do the opposite here in order to explore what latitude observational facts give to theories" (Alfvén, 1979, 27). As we have mentioned, these toy models have never been developed as genuine models of our universe, so we do not know whether the prevailing attitude toward them is fully justified.

The fact that these essentially non-cosmological explanations of the CMB origin often require special physical conditions or special kinds of physical objects (be it unresolved sources, Population III stars with particular properties, or dust grains with atypical shapes and electric properties) stands out as a defining feature of these models and must be evaluated in its own right. In practice, the problem of developing cosmological models on the one hand, and that of considering special physical conditions on the other have been typically conflated. Therefore, the postulation of the existence of certain types of dust in galaxies and intergalactic media may appear to be an essentially cosmological problem engendered by a comprehensive cosmological concept like the material formation of classical stationary-state theory. However, phenomena clearly have local astrophysical properties and thus require a plausible local astrophysical explanation.

All these nonstandard yet moderate cosmological models had varying degrees of difficulty explaining the origin of the CMB. Attempts to assimilate the CMB results into these models (Puget & Heyvaerts, 1980) were ongoing until fairly recently. Like so many other authors developing other interpretations, Jean-Loup Puget and Jean Heyvaerts were directly motivated by Woody and Richards' (1979) results, explicitly mentioning this in the opening and the summary of their paper. They focused on explaining the apparent substantial deviation of the CMB from the blackbody spectrum these results indicated, arguing this accorded naturally with the Population III model. *COBE* 1990 measurements corrected the values and reduced the deviation below 1% and thus decisively refuted this argument for the Population III model. Anthony Aguirre (1999, 2000), however, thought the non-primordial CMB was formed by Population III at high redshifts ($z \sim 100$), and those moderate temperatures could plausibly be thermalized within the current observable limits, providing enough conveniently shaped dust. Aguirre also offered a detailed account of nucleosynthesis in Tepid early conditions, supposedly in tune with the observed quantities of basic elements.

Overall, for most of the past half-century, these alternatives have received a certain degree of support and are considered viable competitors by at least some cosmologists. This is especially true of the period after the great cosmological controversy in the late 1960s. And even in the last quarter of century, variants of nonstandard Big Bang themes occasionally appear in the most prominent astrophysics journals (Aguirre, 1999, 2000; Li, 2003; Yi-Fu, Easson, & Brandenberger, 2012). Unlike attempts in preceding decades, these are not necessarily motivated by considerations immediately related to the CMB.

Crucially, however, the discovery of acoustic peaks by *WMAP* (and some of the high-sensitivity balloon experiments, like *BOOMERANG*; Crill et al., 2003) essentially refuted these accounts.[29] The acoustic peaks detectible in the CMB, predicted by Peebles and Yu (1970), appeared during the early expansion of ordinary matter and radiation due to the perturbations in their density; these perturbations propagated like sound waves (the speed of propagating sound at the time was close to 60% of the speed of light). Now, as there is no early thermal plasma in these models, no acoustic waves can propagate through plasma, and thus, there can be no acoustic peaks in the angular spectrum of anisotropies. In fact, this is clearly a textbook instance of falsification of a hypothesis in recent scientific practice; a set of unique and causally inevitable predictions turned out to be in gross conflict (formally, many hundreds of standard deviations) with outstandingly precise experimental results. Although, generally speaking, the underdetermination of theories and hypotheses by evidence is not transient in cosmology, there are eventually clear refutations. We elaborate on those in Chapter 12; they range from observations that significantly predate the discovery of CMB itself (e.g., the redshift–magnitude relation established by Hubble) to extremely sophisticated modern triumphs of observational astronomy (e.g., the measurement of CMB temperature at distant galaxy clusters using the Sunyaev–Zel'dovich effect).

Before we move on to the next set of alternative accounts, we should note three points relevant to moderate alternatives. First, we should keep in mind that as moderate unorthodoxies have it, a non-negligible fraction of the observed CMB energy density originated in the primordial fireball, and alternatives of the CMB origins may concern only the rest of the radiation. This partition of the background radiation origin is addressed by several models we discuss later.

Second, the sources of the CMB photons could be predominantly accommodated within a primordial origin picture, rather than by scattering processes at the single, brief, all-encompassing recombination epoch. If we make this assumption, there is no a priori reason to expect a perfect fit of these two physically distinct radiation components. As a result, distortions and deviations from the perfect primordial radiating blackbody appear in models relying on this scenario. There are two basic types of distortions: 1) comptonization (or Sunyaev–Zel'dovich) distortions measured by parameter y; and 2) chemical potential (or Bose–Einstein) distortions (see Box 11.1).

Third, the appearance in schemes of the nondesired non-primordial origin of part of the CMB photons is designed to account for completely different observations or theoretical presuppositions. Such is the case, for example, in Nickolay Y. Gnedin and Jeremiah P. Ostriker's (1992) high baryonic density universe. This model was motivated, in part, by a desire to solve the well-known problem of

---

Box 11.1

The "minimal" comptonization parameter y describes a redistribution of photons in reference to the blackbody spectrum, usually via Compton scattering. Yet this value could be generalized to any form of energy release that does not result in the higher temperature blackbody, but in spectral distortion. This becomes another observable property of the CMB, fitting quite naturally into an interpretation within the standard model, as Sunyaev and Zel'dovich's (1980) account suggested, but it could be problematic for some of the unorthodoxies, for example, the model of Gnedin and Ostriker (1992), discussed in Chapter 17. The value of the parameters is usually thought to be zero, in accordance with the limits set by *COBE* (or is unobservable as inferred by Smoot et al., 1992). The comptonization y-parameter, however, is acknowledged as an important measure within the standard CMB interpretation and is the key to understanding subsequent, astrophysical distortions, such as the Sunyaev–Zel'dovich effect in rich clusters of galaxies or the effects of dark energy on small-scale CMB anisotropies. The reference value for the present discussion is again the *COBE* value (Mather et al., 1994); this is also important for later discussions (keep in mind that the condition for significant scattering is usually given as $y \simeq \frac{1}{4}$; cf. Lightman & Rybicki, 1979). While this huge topic is beyond the scope of the present study, the historical process leading to consensus on these deviations is of considerable interest and has not been studied by historians of cosmology so far.

dark matter in galaxies. The authors took great pains to show that the observed smoothness of the CMB is consistent with the primordial origin in the framework of their model in the same manner as in the standard low baryonic density models; the only outcome of the very massive Population III objects they postulated is a Sunyaev–Zel'dovich distortion parameter $y \sim 10^{-4}$. While this can be considered falsified in the aftermath of *WMAP* and *Planck*, it is important to understand that this was clarified only in the last ~20 years. The impression given by streamlined textbook accounts is misleading.

12

Models with unresolved sources

In the early days after the discovery of the CMB, it was technically impossible to observe its "fine structure." In other words, all observations had low angular resolution. The current impressive observational precision of *WMAP* and *Planck* missions inevitably clouds our expectations of what the observational database of the CMB looked like in the late 1960s and throughout the 1970s, that is, in the standard orthodoxy's formative years. It is telling, however, that the so-called dipole anisotropy caused by the motion of the observer, that is, the motion of our solar system relative to the background, was controversial until the experiment of George Smoot and his coworkers with a differential radiometer mounted on a U-2 spy plane (Smoot, Gorenstein, & Muller, 1977; Gorenstein & Smoot, 1981) (see Figure 7.1).

One idea that unsurprisingly came up because of this limitation was that what appears to be a truly diffuse emission, more or less uniform over the entire sky, is actually composed of numerous discrete sources that existing observational techniques fail to resolve. This sort of dilemma emerged with respect to the X-ray background (Fabian & Barcons, 1992), and it still persists today. In a wider context of the history of science, what has been regarded as a *continuum* has often turned out to be better described as a *discretum*, and vice versa. Controversies surrounding corpuscular versus wave theories of light in the eighteenth and nineteenth centuries are well studied by historians of physics (e.g., Whittaker, 1989). Whether "spiral nebulae" are continuous (gaseous) sources of light or ensembles of billions of stars (i.e., discrete sources) similar to the Milky Way has triggered massive controversy in astronomy from Hevelius to Hubble, via memorable moments such as the Kapteyn universe, van Maanen's erroneous observations of spiral arm rotation, the explosion of S Andromedae, and the Shapley–Curtis "Great Debate" of April 26, 1920. Therefore, the controversy of continuous versus discrete origins of the CMB should be seen as natural within the framework of astronomical sciences in general and observational cosmology in particular. In fact, it is remarkable that this particular "sub-controversy" was so brief and so easily resolved.

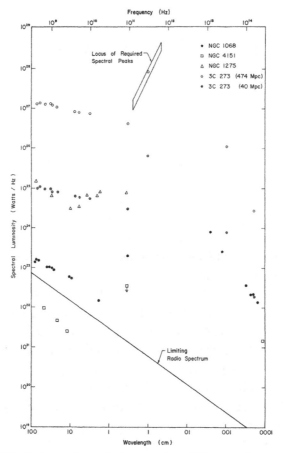

Figure 12.1   Predictions of spectral properties of the cosmic microwave background radiation according to the discrete (yet unresolved) sources model by Wolfe and Burbidge. From Wolfe, A. M., & Burbidge, G. R. (1969). (Republished with the permission of the authors and the publisher.)

Thomas Gold and Franco Pacini (1968) suggested a plausibility argument rather than a serious astrophysical hypothesis. A key question of their approach was whether radio sources might provide enough energy for the CMB. Wolfe and Burbidge (1969, 345) emphasized that there was no convincing explanation for the shape of the observed spectrum or its energy density; energy density was much higher than the integrated value of nonthermal radio sources or regular starlight between galaxies and comparable to the energy density of cosmic rays (Wolfe & Burbidge, 1969, 346; see Figure 12.1). This study is obviously of historical interest, as the authors represented both sides of the orthodox-alternative divide: while Arthur "Art" Wolfe (1939–2014) worked entirely within the standard paradigm, lending his name to half of the Sachs–Wolfe effect (see Chapter 7), Geoffrey

Burbidge (1925–2010) was known not only as the coauthor and collaborator of Sir Fred Hoyle on many seminal studies, including the B^2 FH paper essentially creating the field of stellar nucleosynthesis, but also as a radical maverick, rejecting the cosmological origin for QSO redshifts and assorted other cosmological "dogmas" to his final days.

Gold and Pacini (1968) analyzed spectral distribution with respect to the energy spectrum of the observed body, while Wolfe and Burbidge (1969) took a detailed look at the resolving sources for the steady-state and relativistic frameworks. The latter authors developed an expression for total energy density received by the observer (Wolfe and Burbidge, 1969, 347) and predicted that due to the high energy density of the CMB, there should be high-density (unresolved) sources at adequate frequencies, namely radio frequencies (1969, 346). The candidate sources were identified in millimeter and infrared (between 3 mm and 500 μ) range, with Seyfert galaxy NGC 1068 as another possible candidate (1969, 347). They formulated their general hypothesis as follows: "The shape of the received spectrum will therefore depend upon the nature of expansion [redshift], and the shape and time-dependent behavior of each intrinsic spectrum" for both the steady-state and the evolving universe theories (1969, 354). The way to test their account was to look at the observational accounts and search for possible gaps in the CMB spectrum (1969, 354). Accordingly, they perceived the key insight of the analysis was the limit for radio sources with the given density of the CMB:

We have shown that numerous high-intensity sources which peak in the millimetre and submillimetre spectral regions are required if the microwave background is to be explained in this way. In addition, the spectral luminosity of each source must fall off fairly rapidly on both sides of the peak if the resulting background spectrum is to agree with observation. This may not be the case, however, at frequencies much lower than the predicted peaks, i.e., in the radio region. All the possible types of sources have radio components which exhibit an increase in intensity with decreasing frequency (i.e., they have positive spectral indices) rather than the required decrease. (Wolfe and Burbidge, 1969, 362)

In a discussion of tests of alternative models, Gold and Pacini (1968) suggested that for their theory, "[a] decisive test would be the investigation of the spectral behavior of the background at still lower frequencies – should the blackbody interpretation be correct, no change in the slope would be found, while, if the background is due to unresolved sources, the slope would become $k = 3$" (Gold & Pacini, 1968, L117). Similarly, Wolfe and Burbidge (1969) pointed out that the sources would possibly be resolved near a central peak and suggested the need for observational tests to check this possibility. In particular, they counted on the uncertainties about quasars being resolved as favorable to the hypothesis: "If the radio-quiet QSOs have millimeter luminosities like 3C 273 and are as numerous as recent surveys would indicate, they may very well be the desired sources" (Wolfe & Burbidge, 1969, 363). The most important test should be at 350 μm (Wolfe & Burbidge, 1969, 367) – at far infrared (FIR) wavelengths, unobservable from the ground and even from balloons and planes. Their predictions aimed at distinguishing between the steady-state and evolving universe models (Wolfe & Burbidge, 1969, 368).

Critical assessments of their account quickly appeared in print. First, Cyril Hazard and Edwin E. Salpeter (1969) argued the *lower* bound for the density of potential sources exceeds the density of galaxies (Hazard & Salpeter, 1969, L87), while the existing observational evidence sufficiently demonstrated that rare ordinary objects with the desired density (around the density of galaxies) did not exist, making the hypothesis implausible. Second, a year later, M. G. Smith and Partridge (1970) pointed out that the early measurements were obtained using instruments with poor angular resolution, but new measurements showed the number of sources would have to be very high (higher than the number of existing galaxies) to match the desired data (Smith & Partridge, 1970, 737). Fluctuations in the background radiation occurred at 0.0036 K according to the most precise measurements at the time. Moreover, even for borderline redshifts, a very large number of sources would be needed (Smith & Partridge, 1970, 742; diagram).

The same year Moorad Alexanian (1970) suggested the possibility of a cluster of sources producing a superposition to a single source spectrum of the blackbody type, based on the results in nonequilibrium statistics of quantum mechanics. He argued the condition for this state of affairs obtaining was that "the discrete extragalactic sources must be in equilibrium with the evolutionary temperature $T \propto (1+z)$," and the spectral function of such sources must be provided (Alexanian, 1970, 752)

A remarkable early toy model by David W. Sciama (1966) focused on an alternative explanation of the early temperature measurements of excited CN molecules in interstellar clouds. It provided calculations for the systematic parameters that would characterize the unresolved sources in the radio segment of the spectrum of the CMB survey measurements. The model was admittedly highly speculative (Sciama, 1966, 279), but it was considered necessary to explore alternative explanations because of the potential significance of the consequences if the CMB spectrum was indeed proven to be identical to the shape of the blackbody spectrum, as stated by Sciama (Sciama, 1966, 278).

The CMB source hypothesis assumed an energy density across the spectrum of about 1 eV per cm^3. However, it was possible that the energy density was not evenly distributed as assumed, as some sources may contribute unevenly to the overall energy density due to being unresolved. Therefore, Sciama constructed a physically plausible albeit speculative model of radio sources that could produce an overall CMB survey result, where the radio segment of the spectrum did not appear to deviate from the blackbody spectrum shape due to unresolved sources.

The unexpected intensity in the radio segment was attributed to numerous very compact radio sources. These sources had to be numerous, at least 3,000 times more than the known galaxies, and as compact as stars rather than galaxies, in order to achieve adequate radio luminosity. Sciama pointed out that such compact optical galaxies were known at the time, and these sources would have to be located at large distances since they were unresolved. This fairly plausible model, according to Sciama, challenged the orthodox understanding and even provided possible support to the steady-state advocates.

Development of these models culminated at the end of the 1960s (Wolfe & Burbidge, 1969; Alexanian, 1970; Setti, 1970; Smith & Partridge, 1970). The epistemological attitude behind these models was explicit in the case of Smith and Partridge: "We prefer to make a minimum number of general assumptions and, from these, to derive a general relation between the number density of sources and the amplitude of fluctuations in the background temperature" (1970, 738). A similar, yet reticent, epistemological and methodological attitude was implicated in the study of Wolfe and Burbidge (1969), as they pointed out that no feasible alternative explaining both the shape of the spectrum and its energy density currently existed, and possible alternatives should be pursued. For their part, Hazard and Salpeter (1969, L89) conceived alternatives as an epistemically responsible move; they listed various alternative options after their refutation of discrete sources in the classical steady-state theory.

Two points of wider relevance for the history of cosmology should be mentioned here. First, the classical steady-state theory (still fresh in the astronomical community due to the tireless advertising by its originators, especially Hoyle) by its very definition did not have various cop-outs available to evolutionary alternatives: if a particular class of hitherto unknown sources X was postulated for high redshifts, by virtue of the perfect cosmological principle, sources of the same type X *had to be present* at $z = 0$, that is, in the local universe. Obviously, the inverse is also true: if we observe a paucity of a particular type of source in the local universe, it must necessarily be rare (and hence inefficient) in the high-redshift universe. The classical steady-state theory (and not necessarily the revised or quasi-steady-state theory considered in Chapter 20) was therefore vulnerable to falsification if an explanation based on it needed to invoke *ad hoc* categories of sources.

A second and more general point is that people were appreciative – perhaps *too* appreciative – of the possibility of new, hitherto unknown types of astronomical sources lurking in the distant past, and this appreciation was due to contingent facts of history of astronomy. Notably, two spectacular observational (and at least in part serendipitous) discoveries in the preceding years were of QSOs, or quasars, in 1963 and pulsars in 1967. Both were scientific sensations and attracted significant attention from the popular press and wider public as well. The discovery of quasars was worthy of the front cover of *Time* magazine, and the discovery of pulsars won a Nobel Prize win for Anthony Hewish – but not for Dame Jocelyn Bell Burnell, the actual observer. In such an atmosphere, an air of unconceived possibilities with respect to new classes of astronomical sources likely enabled easier postulating of *ad hoc* hypotheses. The subsequent history of astronomy was, in spite of the tremendous successes, much more sparing with dramatic discoveries of new kinds of celestial bodies (see Harwit, 2019).

Assessing the general state of affairs at the time, Smith and Partridge wrote that "[s]ince its discovery four years ago ... the cosmic background radiation in the microwave region of the spectrum has been variously interpreted" as a nod to a wide array of attempts, although they acknowledged that the Hot Big Bang view was "widely accepted" (1970, 737). They used a conservative assumption to calculate the required number of unresolved sources. First, they had the same apparent luminosity. Second, they appeared independent if relevant data were analyzed statistically. Third, for epochs over 10^8 years, they "turned off" at some point in the past.

Pointing out another general feature of the Hot Big Bang, Alexanian stated that "it may be premature to doubt the equilibrium nature of cosmic background radiation" (1970, 745). The first orthodox account developed was the most accepted based on existing observational values, but a result in nonequilibrium statistical quantum mechanics raised substantial doubt. At the same time, however, fluctuations in a system out of equilibrium, whether random or systemic, are more easily detectable. In particular, Alexanian concluded that while *any* model with multiple sources is likely to be in conflict with the CMB observations in the Rayleigh–Jeans regime, only future improved observations would be able to discriminate between these options.

British astrophysicist Michael Rowan-Robinson published a short succinct paper in 1974 (Rowan-Robinson, 1974), proposing a simple model of discrete sources as the origin of the CMB. Rowan-Robinson entered graduate school at the Royal Holloway College, University of London, about the time news of the discovery of Penzias and Wilson shook the world of cosmology; his doctoral advisor was Sir William McCrea, a veteran of the "great controversy" who once was quite sympathetic to the classical steady-state theory (Kragh, 1996). The British astrophysicist's model represents the key points of this family of alternative accounts in the clearest way. The well-written article painted a simple picture using relevant, but still scarce at the time, observational data. A particularly interesting aspect was its argument that the source of the energy of the CMB photons is reduced to well-known astrophysical objects as sources, namely Seyfert and other types of active galaxies. These sources were intensely observed and researched in the early 1970s; Rowan-Robinson was right on target in expecting that these sources – under the general umbrella term of Active Galactic Nuclei (AGN) – were going to play a more prominent role in future astrophysics. However, subsequent observations of the short wavelengths of the CMB spectrum refuted the basic claim of the model in a straightforward manner.

Rowan-Robinson's work significantly contributed to what has become the standard model of AGNs (e.g., Rowan-Robinson, 1977), the spectrum of which is approximated as a simple broken power law. This theory has a clarity in terms of predictions that Rowan-Robinson's model had, since further research of the spectral energy distribution of the sources might well disprove the theory. Speaking of predictions, Rowan-Robinson stated that the spectrum should not behave as the blackbody spectrum at wavelengths below 3mm, which was consistent with upper limits at the time (Rowan-Robinson, 1974, 47p; diagram). He labeled a test of this prediction a "crucial test" (Rowan-Robinson, 1974, 49p). Steep source counts at 1cm wavelength were also to be expected. Two years later, on a related note, Peter Ade, Rowan-Robinson, and Peter E. Clegg noted the lack of precise observation in the millimeter range around the peak; variability of radiation in that range was needed to form any firm judgment on potential discrete sources (Ade, Rowan-Robinson, & Clegg, 1976, 403). It was also necessary to establish the scale of dust thermalization in galaxies (see Chapter 13). Finally, they predicted: "Discrete sources will probably start to dominate the background at wavelengths $< 300 \mu m$" (Ade, Rowan-Robinson, & Clegg, 1976, 408).

The model did not claim the sources are superposed in a way that mimics the spectrum of the blackbody; such "conspiracy" at the source would be both unrealistic and *ad hoc*. Rowan-Robinson, in fact, proposed that the predictions of the standard orthodoxy and his model start to differ at $\log n > 11.3$, with a sharp increase toward larger frequencies. This cutoff line was characteristic for other alternative models as well.

The challenge did not go unanswered. Hazard and Salpeter (1969) offered an early critique of the discrete source models based on a solely statistical point. This criticism was meant to be independent of the physical issue of thermalization. According to their argument, the anisotropies were to be statistically expected for randomly distributed sources in the order of $1/\sqrt{N}$ (N is the number of sources per unit solid angle, or the beam size). When considering the number of galaxies based on the surveys at the time, this number should be about 0.006 or 0.6%. The relative amplitude of the CMB anisotropies was not known, but it was observationally constrained to less than 0.1% (if we exclude the dipole anisotropy, whose amplitude could not be exactly measured at the time, but whose *shape* was unmistakable). Thus, the supposed background discrete sources had to be much more numerous than the number of known galaxies to result in the remarkable uniformity reflected by the CMB. This further means the postulation of the family of the populations of unknown sources solely determined through the supposed origination of the CMB photons is an *ad hoc* hypothesis of entities that should do the explanatory work. Hazard and Salpeter (1969) made a fair assessment, but we should note that this particular instance of this sort of postulation was not a novelty in the history of science. The postulations of phlogiston or ether, and even (nonbaryonic) dark matter and dark energy in the recent history of physics have the same sort of structure.

It may seem initially surprising that Rowan-Robinson's model used Edward A. Milne's special relativistic cosmological model framework. The reason for this, however, was that "if the microwave background is due to sources, the evidence that General Relativity applies on a cosmological scale is not compelling" (Rowan-Robinson, 1974, 46p). In fact, this argument questions the cosmological viewpoint that the author of the argument embraced elsewhere. We should keep in mind that by the early 1970s, the array of experimental tests of General Relativity had grown enormously, even when applications in astrophysics and cosmology are excluded. This was a consequence of the "mini revolution" in studies of relativity in the 1960s, spearheaded by the work of John A. Wheeler and his students and collaborators, as well as Kip Thorne, Kenneth Nordtvedt, and Clifford Will's development of the parameterized post-Newtonian framework to assess theories of gravity (e.g., Thorne & Will, 1971; Will, 1971; Will & Nordtvedt, 1972). This culminated in the publication of the epochal *Gravitation* in 1973 (Misner, Thorne, & Wheeler, 1973). Without going deeper into this complex and involved historical episode, we may conclude that the application of General Relativity in the cosmological context was not optional any more, as it had been, for example, in the 1930s or 1940s.

Only two years before he published his work on the CMB interpretation, Rowan-Robinson (1972) published an editorial comment in *Nature* titled "Steady state obituary?," in which he tried to hammer the last nail in the coffin of the most prominent purportedly nonrelativistic cosmological approach. He and other cosmologists in the early 1970s were convinced that the emerging orthodox explanation of the CMB origin was by far the most convincing but continued to explore its weaknesses and alternatives. Together with researchers like Rees, Ostriker, Carr, and many others working mainly within the orthodox framework, he performed a useful service to cosmology by probing the limits and boundaries of the accepted wisdom. He continued publishing on distortions of the CMB spectrum by dust (Rowan-Robinson, Negroponte, & Silk, 1979) and microwave emission by different classes of extragalactic objects (Ade, Rowan-Robinson, & Clegg, 1976), but these were mostly indirectly relevant to the issue of the origin of the CMB photons.

Naturally, Rowan-Robinson's particular model of discrete sources was disproved soon after its formulation. Ade, Rowan-Robinson, and Clegg (1976) tried to fix the alternative by providing an argument for plausibility in light of new evidence. Reacting to the refutation by more precise observations, they stated, somewhat desperately:

The observational and interpretational uncertainties do not seem to allow discrete source models of the background ... to be decisively rejected, despite the severe limits imposed on such models by the observed lack of small-scale anisotropy in the background; such models are perhaps of more mathematical than physical interest. (Ade, Rowan-Robinson, & Clegg, 1976, 403)

Yet this statement was in the spirit of the initial toy model by Rowan-Robinson (1974), where the number-density of discrete sources not larger than that of galaxies and energy requirement per galaxy was deemed as not implausible, although unlikely. Rowan-Robinson argued the evolution of properties of sources was crucial, thus echoing the theories developing the idea of thermalization by grains. The model never received a critical treatment and refutation in print, but this is not surprising, given the results of Smoot et al. (1987), as well as a new generation of observational projects that included balloon and plane-based experiments indicating the blackbody shape was satisfactory in the wavelength range of 0.1 – 50 cm. An obvious epilogue was provided, of course, by *COBE*, whose ultrahigh-precision measurement of the spectral shape in the *FIRAS* experiment provided decisive and irrevocable proof of the blackbody nature (Mather et al., 1994).

The development of new models of this sort and interest in them ceased soon after 1974. Unlike models that linked the origin of the CMB to a hypothetical Population III of pre- or early-galaxy objects and subsequent thermalization, discrete source models were a narrowly defined astrophysical class. The postulation of microwave sources relied on a straightforward physics, making the verification of the postulation fairly straightforward as well. The explosion of radio astronomical technology in the early 1970s eliminated this possibility.

13

Thermalization by grains, the first wave

Thermalization by cosmic dust may have been an auxiliary hypothesis, but it played a crucial role in some alternative explanations of the CMB. In the mid-1950s, Harvard-based cosmologist David Layzer (1925–2019) had already considered the problem of galaxy and large-scale structure formation, driven by his interest in cosmological and cosmogonical ideas. He presented the key conclusion of this work in his 1968 paper (Layzer, 1968), where he argued a Cold Big Bang is more favorable to the emergence and development of density fluctuations of the required magnitude than the standard Hot Big Bang of George Gamow and collaborators. He also argued that postulating ordinary well-known processes, as an explanation was epistemically virtuous (Layzer, 1968, 99). Note that this was typical of Layzer, who throughout his long career manifested profound interest in philosophical issues, attested to by his subsequent work on the arrow of time (Layzer, 1976) and even his engagement in wider scientific and cultural matters, like the IQ-related controversies (Layzer, 1972).

Following this epistemic argument, Layzer (1968) postulated thermalizing grains similar to current interstellar grains (composed of ices and graphite), yet of a much larger quantity. He also postulated much less helium in the early universe. This was important in light of the new and impressive work on nucleosynthesis inspired by Hoyle (Wagoner, Fowler, & Hoyle, 1967), which Layzer cited, and which contained the (in)famous "coincidence" argument about the comparable amount of energy released in the hypothetical creation of *all* cosmic helium in stars and the energy contained in CMB photons. Thus, the present thermalization would be very small, but the bulk of energy was released early on when the universe was opaque at optical wavelengths. These early sources of energy were quasars and radio sources, and radio sources and cosmic rays emerged through gravitational collapse.[30] Thermalizers are needed to thermalize 30 K radiation (Layzer, 1968, 100). This was a requirement of some other alternative accounts of the CMB as well.

Some models used thermalizing grains to explain the CMB-hypothesized Cold Big Bang conditions and to explain that the mass of the early universe condensed into stars five to ten times more massive than the Sun. In such a scenario, the early massive objects formed heavier elements, as well as grains. The "missing luminosity" is the CMB, processed by early very efficient sources (Layzer & Hively, 1973, 363) through the gravitational collapse of large objects equal to the galactic mass (Layzer & Hively, 1973, 364), and the grains must be "thin, hollow, spherical shells" (Layzer & Hively, 1973, 367). First, solid hydrogen grains, heated up later on, evaporated and left shells of heavier elements, increasing in density as they moved to the surface and came together into shells (Layzer & Hively, 1973, 367). They were metallic and elongated due to electrostatic forces. This was one of the first instances when the detailed physical properties of dust grains came into focus of the project to explain the CMB; this point is elaborated on in following chapters.

The redshift at which the bulk of the CMB energy had been released, and was not tightly constrained in the Layzer and Hively model, allows the contribution of the gravitational collapse of primordial gas clouds, which could occur only at $z \sim 10$, while the massive Population III stars could release the bulk of their energy somewhat earlier, by $z \leq 50$.[31] The interval $50 \geq z \geq 10$ is the best anyone could do in Cold Big Bang models without having further physical insight into the nature of the Population III sources. What Hoyle and others repeatedly touted as an advantage of unorthodox alternatives – outsourcing the explanatory work to better-known stellar astrophysics – turned out to be an explanatory burden. (As our later discussion of the work of Rees and David Eichler shows, the idea of an evolving universe implies that the early epochs were very different from the recent ones; this does not imply, however, that all degrees of difference are equally plausible.)

We now arrive at the key issue of thermalization of Population III starlight (or infrared luminosity from collapsing gas clouds). According to the observations available at the time, the CMB retains its perfect blackbody shape up to wavelengths of about 20 cm. If this is due to the opacity of dust, then the universe must be opaque at wavelengths below about $20(1 + z)^{-1}$, where z is the poorly constrained redshift of the energy origination. This requires huge amounts of intergalactic dust grains, with huge amounts of metals, leading to all sorts of empirical difficulties, including the expected extinction of light from distant quasars and galaxies. (Conversely, if we think about the set of observed quasars and galaxies as the set of those bright enough objects to be seen in spite of the extinction, the implication is that there is a huge number of these faint sources, making the total amount of matter in the universe, Ω_m, impossibly large.) If grains are roughly spherical with the standard values for dielectric constants, then the fraction of total cosmological mass locked in grains must be enormous: $\Omega_{dust} > 0.06$. This is about an order of magnitude larger than the amount of matter in all luminous stars and gas. It is not realistic today (see the discussion of the high-baryon universe of Gnedin and Ostriker in Chapter 17), but it was not realistic from the point of view of observational astronomy in the 1970s either.

In addition to the spherical grains, Layzer and Hively (1973) discussed what they called a somewhat bizarre case: grains in the shape of hollow spheres (i.e., spherical shells). They offered a plausible qualitative model of the formation of the strange grains: the formation of a carbon and silicate shell around an ice core, followed by sublimation of ice through the porous shell. Layzer and Hively mentioned, but only in passing (and credit the comment to a private communication from Edward M. Purcell, the Nobel Prize winner in 1952 for the discovery of nuclear magnetic resonance and a pioneer of the 21 cm-line radio astronomy), the possibility that the grains may be elongated. This idea of this particular form of thermalizing agents gained popularity in the work of Hoyle and Chandra Wickramasinghe (1967).[32] As these researchers pointed out, the strangely shaped grains could reduce the total mass of dust and metals required. Yet not only is the reduced mass still too large for realistic assessments, but it takes exotic methods to produce such strange grain shapes. Wright (1982) drew this conclusion, first refuting the possibility of spherical dust grains thermalizing. Then, in light of the apparent distortions in the blackbody spectrum suggested by the new results of Woody and Richards (1979), he proposed thermalization with the strange needle shaped grains was adequate. Moreover, along with the strange shape, it was necessary that the grains were not presently emitting (Wright, 1982, 406), and this would make the search for the evidence of the dust grains a difficult task. Possible evidence of elongated whiskers was found in micrometeorites collected in the atmosphere that could have started as interstellar dust; some suggested the abundance of elongated needles in the interstellar dust possibly reflected a long wavelength spectrum in the CMB.

Layzer and Hively's model *requires* that $\Omega_b \approx \Omega_m \approx 1$. Although it does not explicitly exclude any form of nonbaryonic matter, it obviates the necessity for it. Although the total matter density can, in principle, be significantly higher than the critical density, thus enabling a significant quantity of metals to be condensed in thermalizing grains, it is not necessary (and would be problematic for other reasons). However, values for Ω_m much smaller than unity (like the values usually obtained in observational surveys) are strongly excluded, as the small fraction of baryons leads to a still smaller fraction of metals; hence, the density of the thermalizing grains is too small. This is an important prediction, especially because a strong revival of interest in large quantities of dark matter in both the Western world and the Soviet Union came somewhat later than the Layzer and Hively study (Einasto, Kaasik, & Saar, 1974; Ostriker, Peebles, & Yahil, 1974). Layzer and Hively (1973) predicted that quasars at $z \geq 3$ would not be visible due to extinction by thermalizing grains, not due to intrinsic fainting of radiation out of the objects. And the universe (mostly the intergalactic medium) should be opaque to this primary radiation at longer wavelengths, just after the Planck peak of the CMB spectrum, while early matter was condensed in large structures of galactic mass.

David Layzer returned to the topic of Cold Big Bang in the last decade of the twentieth century in a short paper on large-scale structure formation (Layzer, 1992), after the *COBE* data became available. In this study, he reemphasized the advantages of a low photon-to-baryon ratio from the point of view of the growth of density perturbations and in contrast to the then timely HDM versus CDM controversy. Clearly, he had not abandoned all hope of finding a method of efficient thermalization to account for non-primordial CMB origin, despite the *COBE* results. Contrary to the inflationary "paradigm," although they were the result of early fluctuations, he argued that the proto-clusters and protogalaxies evolved distinctly and were thus characterized by different energy properties and were different topologically (Layzer, 1992, L7). Yet the key point of the paper is that isotropy data do not necessarily falsify the Cold Big Bang models since only small-scale anisotropies would result from gravitational clustering of early structures (Layzer, 1992, L6–L7). This assertion should be evaluated in the context of the difficulties theoretical cosmologists were having modeling the transition between linear and nonlinear regimes of growth of gravitational perturbations (e.g., Bond, Kofman, & Pogosyan, 1996).

Finally, in their alternative account of the CMB and thermalization, Alfvén and Asoka Mendis (1977) hypothesized the following:

"The surface of last-scattering" of the observed microwave background radiation corresponds to the distribution of dust in galaxies or protogalaxies with a temperature $T_{dust} \approx 110$ K at the epoch corresponding to $z \sim 40$, and not to a plasma of temperature $\gtrsim 3,000K$ at an earlier epoch ($Z \gtrsim 1,000$), as given by the canonical model of big bang cosmologies. (Alfvén and Mendis, 1977, 698)

Thus, although they postulated a Cold Big Bang, the coupling of radiation and dust occurred in an epoch earlier than postulated in previous attempts. They did not postulate unobserved intergalactic dust but took the known dust distribution in our Galaxy as the input data (Alfvén & Mendis, 1977, 698). We should not conflate the hypothesis of Alfvén and Mendis with the Swedish Nobelist's long-standing interest in cosmologies *without* any Big Bang (e.g., Alfvén, 1979), to which we will return.

14

Primordial chaos

In an early paper, Sir Martin Rees (1972) proposed an additional source of energy for the CMB: the dissipation of primordial chaotic fluctuations. This is not a "true" thermalization model, so we consider it separately from the later model of Rees (1978).

We knew much less about the primordial spectrum of gravitational perturbations (or the "power spectrum") in the 1970s than we know today, so various scenarios could be concocted then. A popular category fell under the umbrella term "chaotic cosmology," although the attribute "chaotic" was not, obviously, linked to issues of predictability and computability as it is now, especially in nonlinear dynamics, celestial mechanics, and many applied sciences. In an important sense, the meaning of "chaos" was antithetical to the one in widespread use now. Primordial chaos served contemporary cosmologists mostly as an *ansatz* whose role was to remove the dependence on initial conditions and to show that various kinds of initial conditions lead essentially to the same universe at later epochs (see, e.g., Misner, 1968). That is very different from the notion that chaos is the property of systems highly sensitive to initial conditions, and that a small difference in the initial conditions leads to exponential divergences in the subsequent evolution (e.g., Strogatz, 2001). This semantic switch should be kept in mind when we study the chaotic cosmology program and other locutions oft-used in 1960s and 1970s.

The first question driving Rees to formulate his 1972 model was the discrepancy between the present inhomogeneity of the universe and the early isotropy reflected in the CMB. Rees identified "the widely accepted version of the 'Hot Big Bang'" (1972, 1669). Yet he characterized the generally accepted framework as a "controversial 'Hot Big Bang' cosmology" because "there is no firm evidence that, on scales up to the size of clusters of galaxies, the universe was ever any smoother than it is today" (1972, 1669). He deemed it desirable that current inhomogeneity follows from theory, rather than be merely coincidental (1972, 1671). Otherwise stated, a theoretical "prediction" of the CMB temperature was needed. The second question was the "unpalatable" assumption of expansion of different parts of the universe with the same curvature and entropy, without causal contact between them (Rees, 1972, 1670). This later became known as the "horizon problem." The chaotic universe hypothesis addressed these two quandaries.

At the same time but independently, the great Soviet/Belarusian physicist Yakov Borisovich Zel'dovich (1914–1987) asked the same question about the current inhomogeneity of the universe and offered a somewhat different solution along the same general lines of looking at the early universe conditions (Zel'dovich, 1972). A key assumption of the Hot Big Bang model that needed an explanation or at least a plausible elaboration was that "initial fluctuations of baryon density are corresponding fluctuations in the metric" (Zel'dovich, 1972, 1P). This key correspondence was accidental, so a model explaining it would have an epistemological edge. Zel'dovich argued the current inhomogeneity resulted from the growth of small early perturbations, which, in turn, depended on the ratio of baryons and antibaryons. The strength of this argument was ultimately a matter of the development of the theoretical details, but his epistemological attitude was explicit: "No apriori preference can be given to small or big perturbation theories – the analysis of observations is the unique approach to the problem" (Zel'dovich, 1972, 3P).

Thus, Rees and Zel'dovich had a similar general epistemological motivation for constructing these two alternatives; both thought the issues of the current inhomogeneity and the isotropy of the CMB had to be resolved by an adequate theoretical account that was ultimately backed by appropriate evidence, not left as mere postulation. They saw their respective theories as first shots in that direction. Their theories were also worked out at the same level: they were modifications of the Hot Big Bang with very similar physical details.

This historical "coincidence" is another example of the abject failure of simplistic social-constructivist notions about the sociocultural determination of the *content* of scientific theories. The conditions in which Rees and Zel'dovich worked could not have been more different in the sociocultural sense. Even the cosmological background was profoundly different, especially as the steady-state theory was never taken seriously in the Soviet Union and other communist countries, in sharp contrast to the West, in particular in the United Kingdom (Kragh, 1996). And yet the substantive content of their work on the early universe was quite similar.

To return to the notion of the chaos in these theories, Rees used primordial chaos in operational terms, as a placeholder for initial inhomogeneities and anisotropies whose smoothing released the required CMB energy. He pointed out that in the standard picture, the redshift of the decoupling of matter from radiation and the redshift of recombination must be very close; he regarded this as an unnecessary explanatory burden.[33] Rees showed how primordial irregularities of different mass scales could provide the continuous input of heat at the expense of gravitational potential and kinetic energy, until the size of the horizon grew sufficiently to encompass the region of the universe we consider to be "sufficiently" homogeneous.

An obvious advantage of pushing the origin of CMB energy further into the past is that higher densities make thermalization easier, *ceteris paribus*. Rees (1972) argued dissipated energy (at $z \geq 10^4$) in the early chaotic universe would have emitted the CMB at 3 K. Moreover, the thermal spectrum "was generated at an epoch when the matter was capable of thermalizing it" (Rees, 1972, 1669), but radiation cannot be ascribed to the high entropy of initial conditions: "At sufficiently early times, the plasma would be dense enough to thermalize the radiation at a temperature T" (Rees, 1972, 1669–1670).

Zel'dovich (1972) made two main points: first, the damping of short acoustic waves during the early expansion resulted in relaxation of thermodynamic equilibrium with high specific entropy per baryon (1972, 1P); second, only baryons and anti-baryons were initially present due to thermal excitations. There was an asymmetry due to density and metric fluctuation, and a particular amplitude of metric fluctuation implies adequate (observed) entropy, chemical composition, baryon/antibaryon ratio, and so on. Rees openly admitted his proposal was rough and speculative: "We leave the physical details of the dissipative processes as an open question" (Rees, 1978, 1670). As he pointed out, the energy density spectrum was still unknown in 1972.

Irrespective of the source, thermalization is easier at early epochs, *ceteris paribus*, as optical depths are huge, even within the context of the open baryonic universes used by Rees ($\Omega \approx \Omega_b < 1$). The degree of "clumpiness" is an important theoretical parameter fixing the smallest redshift at which the thermalization may occur. Parenthetically, this is the only alternative model for the CMB origin that postulates the redshift of origin *higher* than in the standard pictures: as we mentioned previously, Sir Martin Rees estimated that the bulk of energy was released and thermalized by $z \sim 10^4$.

This idea was further tackled by an Israeli astrophysicist, David Eichler (then at the University of Chicago), in a short paper discussing the role of entropy fluctuations in the early universe prior to formation of the earliest visible structures (Eichler, 1977). Eichler added several interesting elements to Rees's picture, notably that the dissipation of turbulence through shock waves is the most important mechanism for the input of heat. Dissipation proceeds until the low mass of $M \sim 10^6 M_\odot$ is reached. He connected this to the formation of the first globular clusters, the oldest existing relics of structure formation. Eichler argued the early adiabatic acoustic fluctuations provided the required scattering of energy prior to recombination (a white noise spectrum of entropy fluctuations) and left a residual density contrast that resulted in the scattering of galaxies. Dissipation occurred at a particular early epoch (Eichler, 1977, 580). Eichler acknowledged Peebles's criticism, noting there was a considerable gravitational influence early on (Eichler, 1977, 581). In effect, he intended to improve Rees's points by providing the details and also offering a critique.

The hypothesis of Rees and Eichler could, therefore, be regarded as a "secondary unorthodoxy" following the primary one of chaotic cosmologies. Subsequent events leading to the demise of the chaotic cosmology program (see, for instance, the account in Peebles, 1993) obviated the need for further research along these lines, even though some important explanatory concerns reappeared, disguised as modern inflationary theories on the one hand, and as the theoretical interpretation of acoustic and later peaks in the map of small-scale CMB anisotropies on the other.

Rashid Sunyaev and Zel'dovich (1980) updated Zel'dovich's idea that some early perturbations in the CMB dissipated over time. Their work was prompted in part by the erroneous measurements of Woody and Richards (1980) that indicated up to a 20% deviation of the CMB spectrum from the blackbody spectrum curve. As we have mentioned, these measurements prompted a new wave of alternative interpretations until the *COBE* satellite measurements reduced the deviations to less than 1% with superior detecting conditions and techniques. In their long technical paper, Sunyaev and Zel'dovich (1980) outlined the conditions under which these early fluctuations would appear and suggested the sort of evidence we could expect. The idea rested on two assumptions. First, radiation energy density and rest-mass energy density increased differently with different redshifts, and they equalized only at a very narrow interval of values, so the domain of fluctuations due to this difference was vast. Second, the CMB spectrum was strongly affected by the release of energy even at small redshifts. This valid concern subsequently became a new field of study, featuring research on *non-intrinsic* small-scale anisotropies in the CMB map, like those produced by the eponymous Sunyaev–Zel'dovich effect (scattering of CMB photons of hot plasma filling the clusters of galaxies) or the Rees–Sciama effect (streaming of CMB photons through time-dependent gravitational potentials during the assembly of galaxies and other structures). Sunyaev and Zel'dovich developed various scenarios of recombination in the Hot Big Bang given these two assumptions and then looked at the distortions of the CMB spectrum in alternative Tepid Big Bang scenarios (they also considered the possibility of observing clusters of hot intergalactic gas in protogalaxies).

Rees's (1972) second critique of the Hot Big Bang addressed thermal equilibration without the causal contact of different parts of the expanding universe – the theory of cosmological inflation emerged in the 1980s to a large degree in order to address this. Thus, what seemed to be distant segments without any contact during equilibration were, in fact, when inflation is counted in, the segments that were separated in a vastly accelerated manner. Now, we should reemphasize a point we stated in Chapter 2: although from both historical and epistemological points of view, cosmological inflation is not a logically necessary part of the standard Hot Big Bang cosmology, it has become a welcome and very fruitful extension of the standard model in the last 40+ years. In fact, the fiercest opponent of the Hot Big Bang model, Sir Fred Hoyle, not only suggested that this has been the only real achievement of modern cosmology, as it provides the explanation of the "mechanism" of expansion, but also argued the proposal of the theory by Alan Guth in 1981 was structurally very similar to Hoyle and Narlikar's C-field theory (we turn to this in due course), as other cosmologists had noticed (Gregory, 2005, 324). Hoyle also thought the two theories were epistemologically quite close; both "begin by identifying new weaknesses of the existing paradigm and then ... show how the proposed new idea removes them" (Hoyle, Burbidge, & Narkilar, 1999, 93).

In particular, *WMAP* and *Planck* have shown that particular properties of the CMB anisotropies map are well accounted for by the inflationary framework (e.g., Komatsu et al., 2011). Theoretical quibbles notwithstanding (e.g., Ijjas, Steinhardt, & Loeb, 2014; Linde, 2014), the inflationary paradigm has brought us much closer to understanding initial conditions of the structure formation process than any other theory. It also promises tighter integration with other branches of fundamental physics, including conformal quantum field theory and string theory.

15

Early intergalactic medium, massive Population III objects, and the large-numbers hypothesis

In 1978, Rees returned to models of a non-primordial CMB origin with a modernized version of Layzer and Hively's hypothesis (Rees, 1978). The motivation for this particular Population III model was, however, different than the motivation for his initial model. The main problem Rees tackled was the origin of the above-discussed photon-to-baryon ratio η. He found the situation in which η could be derived from astrophysical processes and constants preferable to the one in which it had an obscure, albeit cosmological, origin. It seems obvious that such motivation was deeply intertwined with the controversy over fine-tuning physical constants and cosmological parameters for habitability (e.g., Barrow & Tipler, 1986; Barnes, 2012). On the one hand, the issue of initial entropy of the universe was closely related to many of the deepest problems in fundamental physics and cosmology, notably the origin of the ubiquitous thermodynamical asymmetry, often dubbed the arrow of time. On the other hand, the universal expansion discovered by Hubble and Humason established the cosmological asymmetry or the cosmological arrow of time. As mentioned above, David Layzer argued for the idea that thermodynamical and cosmological arrows of time are not only related, but the thermodynamical arrow is determined by the ever-larger configuration space created by the Hubble expansion (Layzer, 1976; see also Davies, 1977; Frautschi, 1982). Since the existence of the local thermodynamical arrow seems to be a prerequisite for the evolution of life and intelligent observers, there is an additional layer of philosophical importance: observation-selection effects, encompassed by the weak anthropic principle of Brandon Carter (1974). We discuss the connection between interpretative hypotheses for the CMB and the anthropic reasoning in Chapter 29.

When he was focusing on the epoch of CMB origination, Rees wrote a very influential paper with his friend and frequent collaborator Bernard J. Carr on fine-tuning and anthropic reasoning generally (Carr & Rees, 1979). The paper massively boosted the interest in Carr's ideas. Carr was also involved in research on the CMB origin, especially the role of Population III objects in supplying energy to the background – thus, we see how tightly interconnected these topics (or *themes*; see Holton, 1988) were.

Now, one of Rees's main motivations for tackling this issue (and most issues in his long and distinguished career) was simplicity. In his own words: "Here a possible non-primordial origin of the microwave background is outlined that seems less contrived than other such schemes" (Rees, 1978, 35). He was as explicit about his epistemological attitude as he had been in his paper six years earlier: "The very early Universe thus seems a topic for conjecture rather than consensus while such obscurity veils the first 3 Myr" (1978, 37).

In pursuing this argument, Rees was modifying the alternatives that accounted for cosmic entropy (since thermalization by helium and hydrogen is not sufficient in existing alternatives). While previous authors were ambivalent on the issue of character and location of hypothetical Population III stars producing the photons to be thermalized, Rees explicitly invoked *pre-galactic* stellar population. Clearly, and in hindsight, this is implausible in view of our understanding of physical preconditions for star formation in later epochs: the density of baryonic matter required for both gravitational collapse and radiative cooling in the absence of metals seems to be too low prior to formation of galactic gravitational potential wells. Again, it is possible to assume a different set of preconditions for Population III stars, but it would be *ad hoc* in the worst manner, a practice Hermann Bondi and Hoyle rightly complained about during the "great controversy."

Even more interesting was Rees's open endorsement of the "large number coincidences" of Paul Dirac and others (see Barrow & Tipler, 1986) as consistent – and perhaps even explanatory – support. Since $\eta \sim 10^{10}$, this is similar to the fourth root of Dirac's large number $\eta \sim \sqrt[4]{10^{42}}$. Rees provided a plausibility argument for the appearance of the fourth root, while commenting that "the general scheme is plausible in order-of-magnitude terms" (Rees, 1978, 37).

Paul A. M. Dirac (1902–1984), one of the greatest physicists of all time, had two periods of intense interest in cosmology, the first in 1937–1938, and the second in the 1970s. His original hypothesis, often going under the name of "large-numbers hypothesis" (LNH) or "Dirac's large-numbers hypothesis," became a full-fledged cosmological model in the 1970s, when it attracted much attention for a time. It is not surprising that the CMB played a significant role in those later discussions, trailing into the 1980s and the *COBE* era. Even a cursory comparison of the two episodes of serious research debates about Dirac's LNH reveals to what extent the conventional wisdom that "CMB changed everything" has in fact been vindicated. Here, we briefly review the most relevant points, noting how the CMB was used as a discriminatory tool in what was arguably the second most important unorthodoxy in the history of physical cosmology (after the classical steady-state theory).

For a more historical treatment focused on Dirac's own work, see Kragh (1982, 2011, 2016) and Unzicker (2009).

In an oversimplified sketch, LNH suggests that big dimensionless numbers of the order of 10^{40} and its powers $\left(\left(10^{40}\right)^0 = 1, \left(10^{40}\right)^1 = 10^{40}, \left(10^{40}\right)^2 = 10^{80}, \text{ etc.} \right)$ are connected to the age of the universe in convenient "atomic" units, namely the time required by light to cross the classical electron radius:

$$r_e^{\text{class}} = \frac{e^2}{m_e c^2}.$$

Thus, the age of the universe is:

$$t_o \approx \frac{c H_o^{-1}}{r_e^{\text{class}}} \approx 10^{40}.$$

The same large dimensionless number corresponds to the ratio of the strengths of the electromagnetic and gravitational forces between an electron and a proton in the atom of hydrogen. From this, the most famous prediction of Dirac's theory follows, namely that the strength of gravity, as expressed by Newton's gravitational "constant" G, decreases with cosmic time:

$$G \propto t^{-1}.$$

Since the total number of nucleons within our cosmological horizon is about $N_n \simeq 10^{80}$, another of Dirac's predictions is (Dirac, 1974, 1979):

$$N_n \propto t^2.$$

In other words, Dirac's theory implies the creation of matter, just like the classical steady-state theory (in whatever manner, either *ex nihilo* or from the universal field with negative energy density[34]). There are further complications, as two sorts of matter creation, additive or multiplicative, have different specific effects on levels such as the Earth and the solar system; those need not concern us here, however.

Yet as strange as Dirac's cosmology is for its local dynamics, globally, it is a Big Bang cosmology roughly similar to other Friedmann universes – as Dirac was quick to point out, it hardly makes sense to speak about the uniquely defined age of the universe otherwise – but with the scale factor growing more slowly with cosmic time: $R(t) \propto t^{1/3}$, instead of, say, $R(t) \propto t^{2/3}$, as in the Einstein–de Sitter model.[35]

While the criticism of the older versions of Dirac's theory focused on local effects, by the 1970s, it was appropriate to place it squarely within the context of the emerging standard Big Bang model, especially in connection with the emerging CMB precision measurements. Variations on Dirac's theme were proposed under the title of *scale-invariant cosmology* by Italian-American physicist Vittorio

Canuto and his collaborators (Canuto and Hsieh, 1977; Canuto & Lodenquai, 1977; Canuto et al., 1977; Adams, 1983). This was most certainly not an outlying or marginal endeavor – some of these papers have attracted moderately persistent attention over the decades, as scientometry suggests.[36]

In an important and highly instructive study, distinguished astrophysicist Gary Steigman (1941–2017), at the time at Yale University, uses the fact that the adiabatic expansion of the universe in the standard Big Bang model preserves the Planckian blackbody spectrum to argue against Dirac's and related models and also against the tired-light models (Steigman, 1978). The reasoning employed is surprisingly simple and accessible, following mostly from the cosmological principle without invoking any particular kind of dynamics. We may define a function α as

$$\alpha\left(t_0,t_*\right) \equiv \frac{N_y\left(t_0\right)}{N_y\left(t_*\right)},$$

where t_0 is the present epoch, and t_* is any fiducial epoch, which may be the epoch of CMB origination as well. For the standard Friedmann models, it is simply $\alpha = 1$ for all times, so the Planck spectral shape is preserved as Tolman first showed in 1934.[37]

Obviously, if the number of photons is not conserved, α will have various values; in general, in Dirac's and related cosmologies, photons are created, and we expect $\alpha > 1$, and possibly $\alpha \gg 1$ if $t_0 \gg t_*$. Now, Steigman first shows that the gray-body spectrum temperature is given as:

$$kT = h\nu_0 \left\{ \ln\left[1 + \frac{e^x - 1}{\alpha\left(t_0,t_*\right)} \right] \right\}^{-1}, \qquad (15.1)$$

where the dimensionless parameter x is defined as $x \equiv \dfrac{h\nu_0}{kT_0}$, T_0 is the present-day CMB temperature, and ν_0 is a fiducial frequency, for which we may take the frequency at the maximum of Planck's distribution curve. From this equation, it is easy to derive a valid relationship for Friedmann universes, namely that

$$T_0 = \frac{T_*}{1 + z_*},$$

if photons were scattered the last time at z_* when the temperature was T_*.

It is also easy to see two classical limits of the spectral distribution: the Rayleigh–Jeans low-frequency limit when $x \ll 1$ and the Wien high-frequency limit $x \gg 1$. Equation (15.1) above gives

$$\frac{T_{R-J}}{T_W} \approx \alpha\left(t_0,t_*\right) = \frac{N_y\left(t_0\right)}{N_y\left(t_*\right)}.$$

If photons are created as in Dirac's model as $N_y \propto t^{2+\varepsilon}$, deviation from the blackbody shape will be enormous unless $t_* \to t_0$ which is clearly unrealistic. In a similar manner, in the tired-light models where the origin of cosmological redshift is not in the expansion of the universe but is due to the energy loss of the scattered photons, it turns out that the ratio of low-frequency to high-frequency temperature is

$$\frac{T_{R-J}}{T_W} \approx \alpha = \left(1 + z_*\right)^3,$$

which obviously cannot be sustained unless $z_* \to 0$, which is again obviously unrealistic. Even if ground-based observations allowed a loophole, this was decisively rebutted by *COBE* results, which put firm limits on the chemical potential to express the nonconservation of photon number (Mather et al., 1994):

$$|\mu| < 3.3 \times 10^{-4}.$$

Together with the blackbody spectral shape with a confidence level expressed by hundreds of standard deviations (Figure 5.1), this result effectively falsifies Dirac-type alternatives.

There was a minor controversy about the possibility of disproving Dirac's theory on the basis of the opacity of the intergalactic medium for radio waves. While Mansfield (1976) argued that the optical depth of ionized intergalactic matter would be too large for extragalactic radio astronomy to be possible, proponents of the varying G theories disagreed. Canuto and Hsieh (1977) gave a reasonable rebuttal in the then less renowned *Astronomy and Astrophysics*, saying Mansfield based his calculation on a confusion between the temperature of radiation in the intergalactic space and the temperature of intergalactic *matter*. The two decoupled at the epoch of CMB origin and have been widely different ever since (especially since intergalactic matter was *re-ionized* in a subsequent epoch $6 < z < 10$). In contrast to Steigman's arguments on the lack of spectral distortion, it seems Dirac's theory and other similar ones could not be brought down by simply considering intergalactic opacity.[38]

Finally, another important feature of Rees's model should be mentioned. It is the involvement of two other processes, (i) reradiation in molecular bands, and (ii) free–free absorption by the ionized intergalactic medium. While the first process requires too many molecules to be efficient, free–free absorption and scattering will create a large optical depth at a longer wavelength, even for small concentrations of free electrons n_e maintained by photoionization (and presumably, though Rees did not dwell on it, shock ionization caused by formation, mass-loss, and explosions of at least some of the pregalactic Population III stars). Rees admitted that "the value of n_e is too model-dependent to permit any firm estimate of how important this process is" (1978, 36). The statement was certainly true in the 1970s, when the state of the intergalactic medium was almost completely *terra incognita*; observations of the Gunn–Peterson effect suggested it was highly ionized, but the exact degree and the representativeness of a few lines of sight observed thus far remained a mystery. Rees gave the redshift of the bulk of the CMB formation as $z \sim 8$ or somewhat larger, much earlier than the earliest stellar/galactic sources we now observe, but appropriate at the time of writing. Note that it is significantly larger than the value assumed by Layzer and Hively (1973) in their model.

The motivation for another alternative akin to Reese's, offered by John D. Barrow (1978), was that inhomogeneity now must mean high anisotropy earlier – it was again the question of why high anisotropy did not result in higher entropy (dissipation) now. He argued that isotropic initial singularity could have resulted in the current state only if stiff form of interacting matter (density equals pressure) took place; it was not perturbed by the kinetic energy of random motions. The CMB emerged as a dissipation of small amplitude (Barrow, 1978, 215). Along similar lines, Rees (1978) stated the Hot Big Bang did not provide reasons for a particular value of η, and accordingly, other explanations were welcome. Since η is used as a proxy measure of the cosmological entropy – or at least it was used this way until supermassive black holes were better accounted for and dark energy was discovered in 1998 (see Egan & Lineweaver, 2010) – it is relevant to the contemporary debates about the arrow of time and the origin of the Second Law of thermodynamics.

Barrow's (1978) alternative was developed as a transcendental argument, to use a philosophical term for arguments of that sort again, and it outlined necessary conditions for a current state. Similarly, Rees outlined a scenario of the emergence of the current inhomogeneous state from isotropic, not chaotic as in his previous version, early state (Rees, 1978, 35). Thus, the main source of the mass in the present universe is pre-galactic dead supermassive stars; they radiated a portion of their mass-energy, and this now appears as the CMB. The hypothesis was conditional on the following evidence: 1) galaxies develop from early irregularities; 2) 80% of the matter in the universe is dark matter in galaxies; 3) the abundance of Fe suggests that pre-galactic nucleosynthesis may have occurred (Rees, 1978, 36). Finally, the physics of thermalization, including thermalizing grains, was uncertain but plausible (Rees, 1978, 37).

Thus, the model hinges on the assumption that early pregalactic sources radiated at the Eddington luminosity, a critical value where outward radiation pressure balances gravity. For accreting black holes or even neutron stars, it occurred naturally, but Population III stars had to be supermassive for the approximation to hold. If it holds, however, a general argument links photon-to-baryon ratio η with the fourth root of Dirac's ubiquitous "large number" 10^{42} multiplied by an array of model-dependent factors of the order unity. Since it is related to the properties of the Eddington luminosity, then, η is not *ad hoc* (or at least not *entirely ad hoc*) in this model but is tightly constrained by "usual" astrophysics.

In his theoretical predictions, Rees (1978) foresaw deviations of the CMB radiation from Planck's law, small-scale isotropy, and anisotropy due to early irregularities similar to the Hot Big Bang predictions, as well as a "missing" galactic mass and discoveries related to its dynamics. Barrow (1978) formulated a prediction in rather tentative terms:

In the light of this circumstantial and indirect evidence it is hoped that further detailed investigations will be made into the status of the Zel'dovich equation of state by high energy physicists and that new theoretical "experiments" may be devised by cosmologists to investigate its implications in the singular environment of the early universe. (Barrow, 1978, 215)

Overall, Rees's model received much more publicity than the one of Layzer and Hively, appearing in popular and semipopular works (e.g., Barrow & Tipler, 1986), and garnering a reasonably high number of citations over time.[39] However, as with other moderate unorthodoxies requiring many luminous Population III sources, its validity hinged on the large baryonic cosmological density and non-standard (or "strange") initial mass function of those early stars. In a very similar model proposed by Shoko Hayakawa (1984), star formation must be strongly suppressed *below* 200 Solar masses (!) to avoid the explosive ejection of metals and the degree of chemical enrichment prohibited by spectroscopic observations. Both Hayakawa and Rees admitted that the presence of nonbaryonic dark matter (and/or dark energy) inflicted a strong blow on any such endeavor.[40]

16

Late thermalization of starlight

In early 1980s, Narayan Chandra Rana (1954–1996), an Indian astrophysicist and a student of Narlikar, published a model with by far the latest thermalization. Rana (1979) offered a detailed theoretical testing of the key detail of a family of the alternatives. In several respects, this is the most radical of the moderate unorthodoxies discussed in this part of the book and serves as a useful theoretical counterpoint to the ideas of Rees, Zel'dovich, and Eichler. It does away with both primordial and Population III (mostly) origins of the CMB photons, proposing instead the thermalization of "normal" starlight at redshifts in the range of $10 \geq z \geq 5$. Since the starlight energy density, as noted above, is much smaller at present than required in this regard, Rana's model requires strong starburst activity at these epochs.

N. C. Rana was a fascinating and tragic figure (Mukhopadhyay & Ray, 2014). Born in extreme poverty in a village in West Bengal, he went on to become a student of the famous relativist Amal Kumar Raychaudhuri and did his PhD thesis under the supervision of Jayant Narlikar at the Tata Institute of Fundamental Research in Mumbai during 1977–1983. The topic of his doctorate was intergalactic dust, whose properties were crucial to his alternative explanation of the CMB. His research was not limited to cosmology, however. He made significant contributions to the study of the chemical evolution of galaxies and the theory of the initial mass function of stars, observed a total solar eclipse, and wrote several astronomy textbooks. He did all this while fighting horrible heart disease (idiopathic hypertrophic sub-aortic stenosis), which forced him to use a pacemaker since his student days. Unfortunately, this intervention was not entirely reliable, and poor health consistently undermined his research and education efforts. He passed away on August 22, 1996 at Pune, two months before his 42nd birthday.

Rana's work on the CMB origin was characteristically innovative. In the 1979 paper, he stated that elongated graphite grains could not effectively thermalize the CMB, except perhaps at decimeter or perhaps centimeter wavelengths. And if scattering and thermalization were sufficiently strong, radio astronomy would have already easily detected the effect (Rana, 1979, 189). Yet largely prompted by the measurements of Woody and Richards (1979) and the inferred substantial deviation of the CMB from the blackbody spectrum, Rana (1980) revisited the whiskers hypothesis and drew a positive conclusion this time, constructing a plausibility argument based on an extensively detailed model of relevant physical properties. He predicted that the pyrolithic graphite whiskers would not strongly affect radio waves (in contrast to his 1979 conclusion based on experiments with graphite) and outlined the conditions that would result in the postulated thermalization, especially the strength of the required sources and their luminosity evolution. In a follow-up piece (Rana, 1981), he again emphasized the new evidence from spectrophotometry provided by Woody and Richards, in light of which he fine-tuned his model. He provided more detailed background conditions of regular sources of energy and thermalization and the scenario of late epoch thermalization by discrete sources. In particular, he considered the helium abundance and luminosity evolution, starting from idealizations to more concrete, realistic details of stellar activity aligned with the background radiation's deviation from the blackbody spectrum.

Generally speaking, Rana's model has several appealing features. First, it does not have the isotropy problem characterizing other models of the CMB origin, including the standard model, nor does it lead to the "horizon problem." As we explained earlier, this problem is one of the most important motivations of modern inflationary models (see Linde, 2008). More specifically, in Rana's model, the angular size of the horizon at, say, $z \sim 8$ (with realistic values of the deceleration parameter q_0) is large enough and may easily comprise the entire sky. This means the isotropy of matter distribution – that is, the cosmological principle of Eddington and Milne – within our visual horizon *implies* the isotropy of the total light contribution. Conversely, the anisotropy signal within the distribution of galaxies will lead to small-scale anisotropies in the CMB map.[41] In addition, this model produces the necessary helium in stars, not in the empirically inaccessible primordial nucleosynthesis. Rana developed a fairly general formalism for thermalization of any cosmological population of sources and the temperature evolution of the resulting radiation. The formalism does not put any extravagant constraints on the nature of the Population III sources: they do not need to be exotic or extravagant, like $M > 200M_\odot$ stars or accreting primordial black holes suggested by other alternative hypotheses. They are just normal metal-poor stars, somewhat more numerous because of starbursts.

In this sense, and seemingly paradoxically, Rana's model is the least assuming of the moderate unorthodoxies. The paradox is resolved when we notice that what is epistemically gained in terms of the nature of sources is lost in terms of the fine-tuning of the whole process. Thus, the failure of Rana's model highlights all fatal flows of the entire project of decoupling the CMB from the Big Bang and the primordial epoch (or from cosmology in a narrow sense). Not only is the thermalizing agency postulated *ad hoc*, as Rana frankly admitted, but a specific chemical composition (graphite) and shape (long needles or whiskers) are required for the thermalization to succeed. The influence of Hoyle, Narlikar, and Wickramasinghe is obvious in this respect, although, to Rana's credit, his mathematical model is fairly general. In hindsight, we see that it tends to overestimate the magnitude of small-angular scale fluctuations in the CMB temperature for about an order of a magnitude, although this is understandable, as the model was constructed a decade before *COBE*. It also favors an open, low-density universe with the best-fit deceleration parameter of $q_0 = 0.12$. This value is certainly better than the $q_0 = 0.5$ used by most cosmologists at the time who worked within the EdS (Einsten–de Sitter) model, but it was falsified after 1998 with the discovery of dark energy (thus making the deceleration parameter strongly negative: $q_0 = -0.6 \pm 0.2$).

17

"An excess in moderation"

High-baryon universe

Gnedin and Ostriker (1992) did not offer a fully developed model the way some physicists did, notably, the steady-state theorists, when they suggested alternatives to the prevailing orthodoxy. Rather, they listed what they deemed deeply unsatisfying assumptions of the Hot Big Bang approach and offered alternative views and predictions for each. In effect, they introduced a few major pieces for assembling alternative models that included their explanation of the CMB sources and spectrum. They were explicit about their approach stating that they had "intentionally not tied the proposed model to any specific cosmological scenario" (Gnedin & Ostriker, 1992, 12). Of course, everything signed by a titan of modern astrophysics, such as Jeremiah P. Ostriker (1937–) or any of his bright graduate students, is bound to attract attention. Such was the case with this rather bold "heretical" study.[42]

An assumption Gnedin and Ostriker criticized is especially striking. They argued dark nonbaryonic matter is one of a few "patches" in the orthodox model "which are quite ad hoc and are accepted only because of our familiarity with them and our basic belief that the underlying standard model is accurate" (Gnedin & Ostriker, 1992, 1). This comment reflects the authors' basic epistemological standpoint that guided their analysis. Furthermore, they argued in favor of the alternative as more "natural and conservative" (Gnedin & Ostriker, 1992, 1) as it relies on plausible regularly explainable astrophysical processes and entities, most importantly their plausible abundance.

At the time of their writing, the baryonic density based on observations suggested substantial uncertainties and possibly a much lower quantity of matter than suggested by the most commonly used EdS cosmological model ($\Omega = \Omega_m = 1$). Most cosmologists were reluctant to invoke cosmological constant/dark energy, and yet both high-redshift and local observations consistently gave $\Omega_m \leq 0.2$, *both* much larger than the available luminous mass in stars and gas *and* much smaller than the theoretically preferred $\Omega = 1$ flat universe (e.g., Lahav, Kaiser, & Hoffman, 1990). This led to the postulation of nonbaryonic matter that decoupled from radiation and baryonic matter. Defending the assumption of an excess of nonbaryonic matter is not only "unnatural" and outside the realm of conservative thinking Gnedin and Ostriker deemed a theoretical virtue in science, but it builds on unobserved postulated entities as a foundation of the orthodox model's key assumption of the amount of matter in the universe. Instead, they built their model, or rather the foundation of a set of models, using "only observed quantities," and this led them to postulate that all the matter that is detected based on its motion – that is, by utilizing laws of dynamics – is all that there is in the universe. They formulated a complicated scheme of early interactions of baryonic matter and plasma that included several regular physical factors (Compton cooling, electron/positron annihilation, ionization, cold and hot radiation distortion, etc.) and early massive black holes of high density, the factors being ordered by their impact (Gnedin & Ostriker, 1992, 5; Table 4). The CMB came out as a product of this complex scheme of astrophysical events involving only baryonic matter and radiation. The resulting values of basic parameters, including the spectral shape and isotropy of the CMB, were deemed underdetermined by observations, including available *COBE* data, requiring further tests that could in principle determine whether the patchwork of the orthodoxy was correct.

While having baryonic matter only is certainly appealing from the point of view of the economy of thought and might indeed be more in agreement with the traditional usage of Occam's Razor in science, the problems arising from the necessity that the extra baryons are *hidden* make any kind of high-baryonic model contrived. Many epistemologists, including Carl G. Hempel, Karl Popper, and Adolf Grünbaum, warned about the impossibility of evaluating explanatory hypotheses in isolation. The network of observations created by modern astronomy had simply grown too large by the 1990s to be accommodated by any simple theoretical maneuver, even if elegant, like the proposal of Gnedin and Ostriker. Extra baryons would have to be hidden *within* galaxies to account for the rotation curves and other local dynamical evidence; thus, invoking massive intergalactic medium as a reservoir for extra baryons fails. In spite of some valiant attempts to do so (e.g., Pfenniger, Combes, & Martinet, 1994), most astrophysicists did not see the way forward with them. The issue was put aside in the "dark energy revolution" of 1998 and the subsequent concordance of cosmological supernovae, CMB anisotropies, and cosmological structure observations (notably baryonic acoustic oscillations; see Seo and Eisenstein, 2003). All of these zeroed in on the small abundance of baryons, $\Omega_b \approx 0.05$. This agreed with the primordial nucleosynthesis constraints and thus provided a brilliant confirmation of what has become the new standard λCDM orthodoxy.

Part V

Radical unorthodoxies: The CMB without the Big Bang

18

Motivations

Who's afraid of the Big (Bad) Bang?

As we described in Chapter 2, the classical steady-state theory of Bondi, Gold, and Hoyle (Bondi & Gold, 1948; Hoyle, 1948) was very much alive at the time of the discovery of the CMB. In his 1928 work *Astronomy and Cosmogony*, James Jeans mentioned the creation of matter hypothesis, and the expanding universe with constant density of matter was floated in a 1937 conversation between Gold and Hoyle (Hoyle, 1948, 372). Helge Kragh traced the origins of steady-state ideas to the early cosmophysical speculations of Svante Arrhenius, Walter Nernst, and Robert Millikan (Kragh, 1996, 143–160, 2015, 192–208).

These ideas were developed by Bondi and Thomas Gold within the framework of the (broadly conceived) Newtonian universes, following the narrow, or "perfect," cosmological principle (Bondi, 1960; O'Raifeartaigh et al., 2014). They required the distribution of material and momentum that was not explained by the underlying dynamics but simply assumed as satisfied at all times. In other words, not only is the universe homogenous and isotropic now, but it remains so at all times. They thought of these ideas as analogous to Einstein's static universe and, hence, as contrasting to Lemaître's (1931) point-source creation. Part of the motivation for developing such a view was a philosophical worry that if we allow laws to change over time in an expanding universe scenario, then the laws we discover with the help of current observations and experiments may not be always valid (Hoyle papers in Gregory, 2005, 40).

Hoyle's version of the classical steady state was rather different. In 1948, he developed a new framework within the GTR, without cosmic constant and with an additional universal scalar field ("creation field" or "C-field"). The cosmological principle is satisfied by the model, in general, and it aims at explaining the expansion and distribution of the material and momentum by local condensations of perpetual matter formation (Hoyle, 1948, 375). The key assumption is that the Riemannian curvature is zero.[43] A radius of an observable universe prevents observation at a point any further than the radius point due to expansion. The total mass of the observable universe is thus independent of time (Hoyle, 1948, 378), while redshift (Hoyle, 1948, 379) and the values of basic parameters of the observable universe are defined accordingly:

It is only through the creation of matter that an expanding universe can be consistent with conservation of mass within the observable universe. ... Extragalactic nebulae are continually passing out of the observable universe, but the total number of nebulae within the observable universe remains approximately constant on account of the formation of new condensations. (Hoyle, 1948, 379–380)

The level of aggregation and size will depend on the age of the galactic group (Hoyle, 1948, 380). Since "[t]he present model has both an infinite future and an infinite past" (Hoyle, 1948, 381), the total entropy does not increase with time. The theory predicted that intergalactic density remains the same irrespective of time (Hoyle, 1948, 378), and this pertains to both gas and hypothetical intergalactic dust grains – a problematic consequence for the theory in later attempts to harmonize it with observed CMB properties. Finally, Hoyle clarified the physical meaning of the suggested formalism: the tensor is not zero since matter is being created, but it has zero momentum. Small deviations from the geodesic should be observed from Earth due to electromagnetic force at the local condensation of matter creation (Hoyle, 1948, 377).

The theory immediately had several distinct problems, mostly with the radio source counts, but also with recently discovered high-redshift objects, QSOs (quasi-stellar objects or quasars), but these obstacles did not seem insurmountable. An excellent monograph by Kragh (1996) showed how the steady-state paradigm overcomes seemingly serious observational refutations leveled at it at the time; for example, the Stebbins–Whitford "effect" (redder colors of distant elliptical galaxies than nearby ones) turned out to be a spurious instrumental artifact. While the discovery of QSOs in 1963 shook the steady-state theory-building – and indeed forced many adherents to abandon it – it was not decisive as the discovery of the CMB turned out to be (in retrospect!). For quite some time after their discovery, QSOs were too exotic even for astronomers, yet, ways were proposed (largely by Geoffrey Burbidge) to incorporate them into the steady-state framework. The framework itself would not be abandoned without a struggle.

Various explanations of the CMB followed in the steady-state tradition. First, the CMB photons were explained as the thermalization of distant discrete sources by some form of cosmic dust grains and, second, as divergent scattering at the "domain boundaries" by particles, mostly electrons, with variable mass. The former explanation is characteristic of the attempts to account for the CMB in both classical and revised steady-state theories, as well as in some more moderate unorthodoxies, like the Cold Big Bang models we discussed earlier. The latter is relevant only for a specific form of the Hoyle–Narlikar conformally invariant cosmological model (see Chapter 19).

A common epistemological argument these authors often voiced explicitly was nicely summarized by Halton Arp and Tom C. Van Flandern (1992) in their identification of three desiderata for a world model that they thought were *not* met by the orthodox model: "(1) It fits and even predicts the observations well. (2) There is no known feasible alternative model. (3) Although there are unsolved puzzles, nothing contradicts the model" (Arp and Van Flandern, 1992, 263). Another very general, and rather nebulous, point (Arp and Van Flandern, 1992, 272) motivating these accounts was that the exceeding smoothness suggests early galaxies did not leave any expected traces.

Other radical unorthodoxies discussed in this part of the book are the plasma cosmologies of Hannes Alfvén (1979) and Eric Lerner (1991) and the closed stationary models of George Ellis, Roy Maartens, and S. D. Nel (1978) and Peter R. Phillips (1994a, 1994b), as well as the time-reversal hypothesis proposed by Paul C. W. Davies (1972).[44] Arguably, these models lack any particularly original points about the CMB phenomenon, a comment that also applies to various fractal/chronometric cosmologies, Dirac's large-number hypothesis, or tired-light models.[45] These models mostly rehash the explanatory mechanisms developed by the steady-state (or Cold Big Bang, in some cases) proponents, notably the thermalization of background sources on dust grains. In particular, the tired-light models claiming that photons lose energy either by interacting with the intergalactic medium or by traveling through a vacuum, that is, "on their own," lead to the claim that the universe is not really expanding. The models have obviously been refuted and are sometimes seen as hallmarks of pseudoscience.[46] This does not mean there are no epistemological lessons to be learnt from them. On the contrary, the appearance of a new generation of cosmological controversies and discontents in the twenty-first century, focusing on cosmological inflation and other allegedly "outlandish" parts of the dominant paradigm (e.g., Disney, 2000; Ijjas, Steinhardt, & Loeb, 2014; López-Corredoira, 2014), forces us to acknowledge some important historical precedents. We discuss these in detail in Part VI.

19

Hoyle–Narlikar theory and the changing masses origin of the CMB

In response to the fall of the classical steady-state theory, during the late 1960s and 1970s, Sir Fred Hoyle and his brilliant student Jayant Vishnu Narlikar developed an array of new theories based on the general idea of large-scale stationarity. Although the new theories were simply variations of the classical steady-state concept, in the view of the authors, they represented a "radical departure" from it (Hoyle and Narlikar, 1966). They centered on extensions of Hoyle's field theory version of the classical steady-state cosmology to produce a conformally invariant cosmology. It is indicative that the other two fathers of the classical steady-state concept, Sir Hermann Bondi and Thomas Gold, publicly disavowed any association with later Hoyle–Narlikar versions.[47] Hoyle thought, however, that new observations would prompt a review of fundamental elements of the Big Bang models and would not result in what he believed was a hasty convergence in the face of ever-more puzzling data. In his 1972 address to the Royal Astronomical Society, he voiced his concerns:

Do we cross a bridge into wholly unfamiliar territory or do we try to remain safely within well-known concepts? For me personally, the exact state of the data at any given moment is less important than the trend of the data. ... Either the bridge must be crossed or one must judge the data of the past five years to be extremely freakish. (Hoyle in Gregory, 2005, 322)

This was simply the latest expression of a general attitude Hoyle shared with his collaborating group, namely that new observations should prompt reconsideration of existing models and ideas, rather than conform to them (Hoyle in Gregory, 2005, 332).

This, of course, is a complaint often heard about *any* scientific orthodoxy, especially if it has become solidified after a protracted struggle. There is no dearth of examples outside the realm of cosmology. Gould (2002) described the process of the solidification of Modern Synthesis in evolutionary biology (although not perhaps in an entirely nonpartisan manner), taking similar jabs at adherents to orthodoxy as prone to seek empirical confirmation instead of reconsideration. We return to this epistemological regularity in the concluding part of the book.

All in all, a completely different set of questions from those asked within the framework of the Hot Big Bang was supposed to be posed, and new ideas were to be floated. As Hoyle and Narlikar commented: "It is possible to establish a connection between galaxies and cosmology – clearly an advantage over the homogeneous theory that dismisses galaxies as local irregularities" (Hoyle & Narlikar, 1966, 162). Similarly, quasars and other anomalous extragalactic radio sources were treated as residues of matter production (Hoyle & Narlikar, 1966, 175). In this line of thinking, the creation of matter does not occur uniformly but in isolated centers; a steady expansion occurs due to the creation of matter itself in the pockets (Hoyle & Narlikar, 1966, 162). Moreover, the creation of matter takes place in the pockets (and not as homogenous creation of baryons and leptons everywhere as steady-state theory would have it) in the neighborhood of massive objects (Hoyle & Narlikar, 1966, 163). In other words, there is another, very high level of violation of the cosmological principle: the creation of new matter is radically inhomogeneous, this time on large spatial scales. In and of itself, this represents a loss of a powerful epistemic tool: if inhomogeneities are not confined to small spatial scale, any particular observed large-scale galaxy distribution feature (superclusters, "great walls," voids, etc.) could, in principle, be "explained" by postulating corresponding initial inhomogeneities. Ironically, this sounds exactly like what classical steady-state theoreticians mocked as false Big Bang "explanations": the universe is as it is because it was as it was.

A question on the back burner that did not address the Hot Big Bang directly was whether the creation rate in the neighborhood of massive objects is likely to be large (Hoyle & Narlikar, 1966, 163–164). Do particles created in the pockets have enough energy to escape to infinity (Hoyle & Narlikar, 1966, 164)? Hoyle and Narlikar discussed the conditions whereby an electron reaches infinity and concluded it escapes from the object "for all energies up to 10^3 GeV" (Hoyle & Narlikar, 1966, 165).

They also speculated the origin of cosmic rays was cosmologically relevant (Hoyle & Narlikar, 1966, 165–167). In fact, the nature of their origin and properties was directly related to a dilemma addressed in the early accounts of the dynamics and properties of intergalactic matter and galaxy formation starting with Sciama (1955), with debates on the exact amount of matter produced by stars that accreted around them and the amount that dissipated and aggregated into hot clouds. How dense were these clouds, and how quickly did they cool, providing the estimated amount of helium in them and the contribution of heating by cosmic rays? This combination of very complex observational and theoretical questions is still not fully resolved.

The fate of Hoyle and Narlikar's theory depended on very specific assumptions of both the helium density in the clouds and the contribution of cosmic rays. The best available account of the density and the cooling process at the time (Gould & Ramsey, 1966) did not favor their view.

The explanation of the CMB origin in the 1972 version of these theories (Hoyle and Narlikar, 1972a, b) is both provocative and intriguing – and it directly hinges on the *dynamics* of conformal gravity as *a theory of gravity*, discussed in detail in a beautifully written paper by Hoyle (1975). Basically, by extending the ideas of Wheeler–Feynman classical direct-particle interaction model from electromagnetism to gravitation, Hoyle and Narlikar created an unorthodox theory of gravity. The key idea was that, while dimensionless quantities are all fixed, we should be able to express any dimensional quantity using particle masses and these dimensionless quantities – and, hence, *translate any dynamic aspects of the evolution of any system into changes in the masses of particles* over spatiotemporal coordinates.

Now, it turns out that the application of Hoyle–Narlikar's gravity to cosmology requires variable masses of particles as functions of their spatial coordinates. If we accept such variation of masses, it is clear that under fairly general conditions, at some point in spacetime, all the masses will become negative. This is not very disturbing, as it is only important to have all the masses in a single causally connected region of the same sign.[48] Thus, we may imagine the universe consisting of two halves, with opposite signs of the masses and opposite signs of the mass field contributing to the action described above. Every result of an observation is a dimensionless number where the sum of dimensionalities of basic parameters leading to the result is zero. Thermalization happens at the zero surface – radiation is strongly absorbed and reemitted. And thus, a normal starlight explains the density of the CMB. By further generalizing this picture, we may get a whole net of aggregates (one of the first multiple-universe schemes in cosmology!) instead of two, as shown in Figure 1 in Hoyle's (1975) paper (see Figure 19.1).

Obviously, the interface between the two halves of the Hoyle–Narlikar universe is the zero-mass surface. Strange physics taking place in this region because of vanishing particle masses may create a simulation of the physical condition near the Big Bang in the Friedmann cosmologies, from the point of view of a distant observer. In particular, the scattering amplitudes tend to infinity, as they are inversely proportional to the masses of scattering particles. This pertains to electrons, which scatter any photons present extremely efficiently. A very large (formally divergent) amount of scattering is bound to produce the exact blackbody spectrum, indistinguishable from what has been observed. In this respect, the zero-mass surface in Hoyle–Narlikar's theory corresponds to the surface of the last scattering in the standard Hot Big Bang cosmology or to a homogeneous

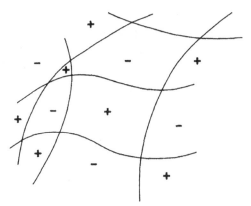

Figure 19.1   Hoyle's multiple-universe schema of spacetime. Hoyle, F. (1975). On the origin of the microwave background. *The Astrophysical Journal, 196,* 661–670. (Reprinted with the permission of the publisher.)

distribution of thermalizing grains in models with the thermalization of distant sources (like the Layzer and Hively model).[49]

Any amount of matter, even a minuscule amount, near the zero-point surface will act as a perfect thermalizer – and if we stick to the cosmological principle of homogeneity and isotropy within each domain (as Hoyle and Narlikar did), there is no reason to expect any dearth of matter near the boundary. Thus, an interesting or bizarre consequence of Hoyle–Narlikar's theory is that all primary anisotropies should be exactly zero. In stark contrast to both the standard model and all other alternative hypotheses for the CMB origin, in this model, the spectrum will always remain a featureless blackbody. This was the reason for Narlikar and Rana's (1980) claim that the Hoyle–Narlikar theory offers a better fit to the CMB – in pre-*COBE* times! – than the standard cosmological model. Narlikar and Rana's argument was an exception to the wave of optimism among creators of alternatives provoked by the apparent distortion in the blackbody spectrum of the CMB of Woody and Richards (1979). Narlikar and Rana (1983) had to demonstrate why the apparent distortion was not a decisive blow to their view; to that end, they appealed to possible systematic errors in the measurements, that later on, with *COBE* measurements, turned out to be large. Of course, the advent of *COBE* and other experiments revealing complex and extremely informative structures in small-scale anisotropies falsified their radical prediction as well (Wright et al., 1994).

Like several other alternatives, Hoyle's 1975 account was driven by the previously discussed puzzle of the uniformity of the CMB, while in the Hot Big Bang, different parts of the universe were never causally connected because of expansion. Hoyle constructed a very abstract theory; it was a far cry from the model built ground-up from physical details. He also explained a seeming coincidence between intensity of the CMB and energy of conversion of hydrogen to helium within galaxies; he did so by introducing the interaction of molecules of deuterium that produced helium and neutron at the "zero-point surface" (Hoyle, 1975).

Unfortunately, Hoyle's paper has received little attention.[50] Obviously, the Big Bang orthodoxy became much stronger after the discovery of the CMB. Other arguments on the nature of gravitational interactions in the conformally invariant cosmology led to the theory and the cosmological superstructure it suggests being considered highly implausible. Despite the relativism of any aesthetic choices in science aside, however, it is our impression that the elegance and subtlety of the Hoyle–Narlikar theory is unmatched by other CMB explanations, whether orthodox or unorthodox.

20

Revised steady state

Narlikar and Wickramasinghe's 1968 theory was an abstract steady-state theory combined with thermalization by dust grains hypothesis and discrete sources hypothesis, and it was developed against the backdrop of still uncertain measurements. These authors argued the measurements do not suggest the CMB radiates as a blackbody at all frequencies, and they pointed out that it does not follow the exact blackbody curve they described at the beginning of their paper (Narlikar & Wickramasinghe, 1968, 1235). Shakeshaft and Webster (1968) critiqued their model arguing the microwave radiation would have to exceed visible radiation by a factor of one hundred, and a very special unlikely distribution of oscillators (at the frequencies of 36.6 GHz to 408 MHz) was required to fit the hypothesis. Although not entirely refuted, the model was deemed very unlikely given the observations of the CMB at the time.

Wickramasinghe et al.'s 1975 paper is an extreme example of a constructive sort of alternative hypothesis. Developed as an alternative to the alternatives of early theories of thermalization by grains, it provided very specific physical details of the properties of graphite whiskers thermalizing the early radiation (their size, electric conductivity, metallic content and its abundance, tensile strengths, shape, mass density, electromagnetic absorption properties), within a certain set of global conditions.

In the early 1990s, the last and most comprehensive instance of the classical steady-state theory was formulated under the name of "quasi–steady-state" (henceforth QSS) theory, sometimes called revised steady-state theory, in a series of papers by Hoyle, Burbidge, and Narlikar, and occasionally Arp, Wickramasinghe, and Sachs, and a few other collaborators. Without going into the wealth of technical details (presented *in extenso* in Hoyle & Burbidge, 1992; Hoyle, Burbidge, & Narlikar, 1993, 1994; Narlikar et al., 2003), we should mention that as in Hoyle's version of the classical steady-state model, the negative energy of the creation field (C-field) transforms into matter with positive energy. However, the creation is not uniform in spacetime but occurs in discrete creation events, the so-called "mini-bangs." In each individual "mini-bang," about $10^{16} M_\odot$ (a characteristic mass of superclusters of galaxies and a mass similar to the "Zel'dovich pancakes") are created in the form of particles with Planck mass $\left(M_{Pl} \sim 10^{-5} \text{ g} \right)$. The distribution of creation events creates the characteristic cellular structure seen in the large galaxy surveys of recent decades.

The motivation for this version of the steady state was multifold, but three goals stand out. First, the authors aimed at eliminating singularities of the relativistic orthodoxy, going a step further than Ellis, Maartens, and Nel (1978) or Phillips (1994a, b), to whom we turn shortly, who retained their own accounts of singularity. Second, Hoyle, Burbidge, and Narlikar (1993, 437) insisted the creation of matter had to be explained as an inherent feature of the laws of nature, not as an accidental occurrence, and postulated initial conditions, direct evidence of which can be obtained in principle, unlike the postulated initial conditions of the Hot Big Bang (Narlikar, 2003, 585). Third, the authors argued strong gravitational fields in galactic nuclei are the sites of nucleosynthesis, and they described the mechanisms to great length. They predicted slight gravitational anisotropies and suggested interferometric test measurements using LIGO (Laser Interferometer Gravitational-Wave Observatory).

The scaling factor in QSS cosmology has an exponential component (as in the classical steady state) and an oscillating component, typically of the following form:[51]

$$R(t) = e^{\frac{t}{t_c}} \left[1 + \varepsilon \cos\left(\frac{2\pi\tau}{Q} \right) \right].$$

Here, t_c and Q are two timescales, one for the conventional Hubble expansion, and the other for the temporal amplitude of mini-bangs, while τ is a function of cosmic time t, which deviates from t only near minimal values of $R(t)$ – that is, near the local "mini-bangs." These epochs are characterized by the creation of new matter, ultimately in gaseous form. Parameter ε has an absolute value less than unity, so the scale factor never actually reaches zero.

Hoyle, Burbidge, and Narlikar's (1993) explanation of the origin of the CMB is essentially the same as the explanations in earlier thermalization propositions, although, of course, there is no truly universal Population III in QSS. Each creation event possesses its own "primordial" stellar population. In accord with the general motivation for the model, Hoyle et al. stated the following in a rather belligerent manner, while rehashing well-known objections:

It is often stated that big bang cosmology explains the microwave background. It does no such thing, of course. Big bang cosmology assumes the microwave background, and it does so in a quite arbitrary way, requiring the baryon-to-photon ratio to be close to 3×10^{-10} without offering a convincing explanation of this number, which could just as well be anything at all. (1993, 443)

The conclusion is somewhat ironic, as Hoyle seemed to have forgotten his own "anthropic" prediction of the ^{12}C level necessary for stellar nucleosynthesis; it makes sense to conclude, as some authors have done (e.g., Barrow & Tipler, 1986) that anthropic reasoning could fix the η ratio as well.

A more novel idea is that thermalization is carried out in two phases, first by carbon whiskers converting starlight into infrared radiation and next by iron whiskers ("needles") producing the observed microwave background. The only component of the integrated starlight that cannot be entirely thermalized is the starlight originating in the last generation of supercluster formation, that is, the last mini bang. As Narlikar et al. pointed out, "These will stand out as inhomogeneities on the overall uniform background" (2003, 2), like the small angular scale anisotropies discovered by *COBE*. These authors even sifted through the available measurements of inhomogeneities in the CMB (Narlikar et al., 2003) in order to compare them with the results of the theory.

Simply stated, however, this cannot work, and Wright et al. (1994) and Wright (2003) pointed to many fatal problems with the account. Notably, the metallic whiskers required for thermalization would cause a huge optical depth of the order of 100 (!) in millimeter wavelengths toward sources located at a redshift of about 2. The fact that we readily observe some such sources among IRAS superluminous galaxies or in the Hubble Deep Field and its follow-ups, for example, argues against the QSS scheme. In addition, the implied power spectrum of perturbations is incompatible with both *COBE* and *WMAP* data, predicting the spectral index of $P(k) \propto k^n$ to be $n = 3$, while the *COBE* value is $n = 1.2 \pm 0.3$, and the *WMAP* nine-year dataset value is $n = 0.972 \pm 0.013$. As mentioned in Chapter 3, this flat spectrum of small-scale anisotropies is one of the major arguments in favor of the standard λCDM model and inflationary setup for the initial conditions. No mechanism suggested thus far could remove those discrepancies, while retaining a remote chance of QSS being adequate.

Two twenty-first-century studies of interest to this family of alternative views (QSS has been modified several times since its proposal by Hoyle, Narlikar, and Burbidge) are those by Li (2003) and Fahr and Zönnchen (2009). Li (2003) returned to the topic of the convenient "needle" shape of intergalactic dust as a necessary thermalization agent, not only for QSS, but also for a host of other alternative models. These improvements to the theory of the extinction of electromagnetic waves are substantial, although still not sufficient to determine whether a convenient form of grains can be found. For instance, Li (2003, 598) refuted the option of infinite cylinders as possible candidates, but left open a possibility of exceedingly thin needles as adequate for thermalization. Despite the rather tentative nature of these proposals, it is instructive to briefly compare this work with the early discussions of dust thermalization cited above (Narlikar

& Wickramasinghe, 1968; Layzer & Hively, 1973; Wickramasinghe et al., 1975; Rowan-Robinson, Negroponte, & Silk, 1979; Rana, 1979, 1980, 1981; Carr, 1981a, 1981b; Bond, Carr, & Hogan, 1991).

After the first comprehensive results from *COBE* were published, some authors (e.g., Carr, Bond, & Hogan, 1991) explicated a basic assumption, whereby a set of numerical models of CMB thermalization by dust grains potentially produced by pre-galactic discrete sources, predicted small anisotropies that *COBE* could detect within the acceptable limits (Carr, Bond, & Hogan, 1991, 446). They introduced the early and late anisotropies, the former due to usual Big Bang developments, and the latter due to scattering by inhomogeneous hot gas (Carr, Bond, & Hogan, 1991, 432), which they argued was as important and powerful. Their expectation was basically that the more detailed *COBE* results would show distortions in range similar to that proposed by Woody and Richards (1981).

Prompted by the incoming results, the Population III sources and thermalization by grains were combined in several accounts in a similar way: the authors focused on the key assumptions of a family of relevant models.

Dielectric dust grains seem to be conclusively rejected now, whatever shape we consider; only speculative conducting ("metallic") elongated grains ("needles") remain even remotely viable. Their tenuous viability is undermined by the fact that the "widely adopted Rayleigh approximation is not applicable to conducting needles capable of supplying high far-IR and microwave opacities" (Li, 2003, 598). No one has proposed an adequate model, causing Li to reach the following conclusion:

Due to the lack of an accurate solution to the absorption properties of slender needles, we model them either in terms of infinite cylinders or according to the antenna theory. It is found that the available intergalactic iron dust, if modelled as infinite cylinders, is not sufficient to produce the large optical depth at long wavelengths required by the observed isotropy and Planckian nature of the CMB. However, the applicability of the antenna theory to exceedingly thin needles of nanometer/micrometer thickness needs to be justified. (Li, 2003, 598)

It seems we are back where we started. By the beginning of the new millennium, the search for thermalizing dust grains had apparently become what Imre Lakatos (1978) called a degenerative research program. This becomes obvious, for instance, in a paper by Hans J. Fahr and Zönnchen (2009). These authors critically reviewed major results of the CMB astrophysics and proposed retreating to what is essentially the QSS framework. They recycled an old argument of Rees (1978) on the dimensional analysis of the baryon-to-photon ratio, this time without emphasizing that fusion/accretion must proceed at the Eddington limit to provide the necessary energy of the CMB. Ironically but fittingly for this rear-guard action, they argued that even the entirely hypothetical metallic whiskers are not enough for thermalization to satisfy modern observational constraints, so they invoked the ancient tired-light yarn (photons lose energy as they interact with matter while traveling through the static universe) as part of the explanation.

All in all, QSS theory comes across as similar to the Cold Big Bang cosmology in its explanations of the CMB origin and must face the same objections as the models of Layzer and Hively, Rees (II), and Rana.[52] Despite the valiant efforts of Hoyle, Narlikar, Wickramasinghe, and others, even *ad hoc* postulated thermalizing agents are incapable of saving the theory.

21

Closed steady-state models

Closed steady-state models were developed after 1965 by theoretical cosmologists with no serious desire to challenge the Big Bang orthodoxy. In contrast, they were all consequent relativists, and "flights of fancy" were remote from their minds. Instead, they wanted to achieve a better understanding of the underlying assumptions of the relativistic cosmological models, the boundaries of their sets of parameters, and their relationship with the empirical data.

Distinguished South African relativist and cosmologist George Francis Rayner Ellis (1939–) was a pioneer of this research. Ellis (1977) introduced the following hypothesis: a static spherically symmetric universe within the general relativistic frame is inhomogeneous, and our Galaxy is close to one of the two centers. The systemic redshift discovered by Hubble and Humason is due to gravitational redshift, not to universal expansion. The CMB is radiating from the second center – its temperature increases with proximity to this center. The singularity at the center is the source of radiation due to circulation of matter, especially light elements (Ellis, 1977, 93). Ellis also defined an explicit refutation criterion: theoretically, the theory is ruled out if a substantial variation of a key parameter (e.g., Hubble constant) is observed (Ellis, 1977, 92).

An explicit epistemic motivation of the theory was to challenge the cosmological principle:

Models of the sort described here have not been considered previously because of the assumption – made at the very beginning, in setting up the standard models – of a principle of uniformity (the cosmological or Copernican principle...). This is assumed for a priori reasons and not tested by observations. However, it is precisely this principle that we wish to call into question. The static inhomogeneous model discussed in this paper shows that the usual unambiguous deduction that the universe is expanding is a consequence of an unverified assumption, namely, the uniformity assumption. The assumption is made because it is believed to be unreasonable that we should be near the centre of the universe. (Ellis, 1977, 92)

Ellis offered anthropic reasoning as a reason for us being close to the center, where the conditions are favorable to life, which is close to the center.[53] He further argued the isotropy of the CMB is not necessarily an argument about our own position or isotropy of the universe (1977, 93).

In a more extended version of the argument Ellis, Maartens, and Nel (1978, 440) stated that the claim about the homogeneity of the universe is an assumption of the Hot Big Bang. Isotropy is directly observable, unlike homogeneity. (This point was, interestingly enough, known by William Herschel in the eighteenth century in the context of the distribution of stars within our stellar system.) Thus, there is – and will be, at least until intergalactic travel is developed – a metaphysical element in our assumption of homogeneity. Finally, the static models bypass the lack of causal communication in a uniformly expanding isotropic universe, so there are no problems with horizons, unlike in relativistic cosmology.

Ellis essentially offered another plausibility argument explicitly pointing out the state of protracted underdetermination in cosmology: "It is not claimed that the universe is actually like this model. What is claimed is that there are no overwhelming arguments that immediately show that such a model could not reproduce all the current observations" (1977, 93). Ellis, Maartens, and Nel (1978) acknowledged the approach was an exploratory model lacking fit with the observed magnitude–redshift relationship, but they thought a version of it should be possible:

Although our investigation suggests that an exactly static inhomogeneous cosmology is not viable, it points out certain interesting features of such cosmologies which might remain in realistic expanding inhomogeneous universe models; in particular, the singularity structure in such universes could be completely different from that in a FRW universe models. (Ellis, Maartens, & Nel, 1978, 440)

They identified a major obstacle to the method: even simple equations of the state cannot be solved analytically (Ellis, Maartens, & Nel, 1978, 451). And they pondered the options in a section titled "Alternatives" (Ellis, Maartens, & Nel, 1978, 455).

Ellis, Maartens, and Nel were even more explicit about the cosmological principle: "Our result constitutes a proof that the universe is expanding, which is not based on a priori assumptions about space-time geometry" (1978, 439). In the conclusion to their paper, they stated: "There may be better methods of proving this result, than those we have given; our contention is that some such proof must be given, for otherwise the statement that the universe is expanding cannot be regarded as an observationally-based statement" (Ellis, Maartens, & Nel, 1978, 461). They basically developed a transcendental argument here or, as they labeled it, a *cosmographic approach*, asking what models it would take to obtain the observed values. The guiding idea was to "work out the observational relations that would be seen in such a space-time by a 'fundamental observer'" (Ellis, Maartens, & Nel, 1978, 441). Thus, "adjusting the space-time geometry to reproduce the observed relations" in the cosmographic approach is contrasted with the "*cosmological* approach" that develops a dynamic model based on Einstein's equations and provides predictions of the observables. In their model, which aims at studying global assumptions like the cosmological principle, the universe is isotropic only around a specific point in space; we are located in close proximity to this point, because of the anthropic selection effect. Following the weak anthropic principle, it is expected that we are located in regions possessing the necessary properties for the creation and evolution of complex biological systems, near the "center" of the universe (it is most natural to use this term for our pole of the manifold, by analogy with a three-sphere). A singularity surrounded by hot matter is located opposite the center, simulating the initial singularity in the Friedmann models. However, in this static model, the singularity is copresent with everything that exists; it does not precede it. Obviously, this makes the model more appealing from an epistemological point of view, although the laws of nature break down at singularities.

In Ellis et al.'s model, the global cosmological singularity is not always inaccessible in the past; it could, in principle, be investigated using the methods and apparatus of modern science. This could be read as a late echo of ideas promoted in the time of the classical steady-state theory, especially Sir Hermann Bondi's endorsement of an extreme Popperian view of falsifiability, and this would discard any considerations of unobservable cosmological phenomena, such as those beyond our particle horizon or before the universe became transparent at recombination.[54]

This copresent singularity is easiest to intuitively understand as "an enclosure" or "a mantle" surrounding the universe. Its major purpose is to be a recycling facility in the global cosmological ecology, as the static nature of the universe makes the recycling of high-entropy matter necessary to explain the obvious and "anthropic" observation that the universe is not in a heat death state. In Ellis et al.'s model, this is achieved by the streaming of high-entropy matter (mainly in the form of heavy elements synthesized in stellar nucleosynthesis) toward the singularity, where it is dissociated and returned to the universe in the form of low-entropy matter, presumably hydrogen atoms or plasma of baryons and leptons. Other than this streaming, which does not change the net mass distribution, there is no systematic motion: all observed redshift is entirely of gravitational origin.

Ironically, although this model is highly nonstandard and was rejected from the start because of its inability to properly account for the redshift–magnitude relations of galaxies known since Hubble, its explanation of the CMB origin is the standard one rehashed. The CMB photons originate in close spatiotemporal proximity to the global singularity, and this proximity is in both Ellis et al.'s model and the standard cosmology, hot and optically thick. The only difference is that we can speak of a continuous production of the CMB in the static model of Ellis et al., while in the standard picture, and all other alternatives considered above, the CMB was produced in a definite past epoch of the universal cosmic time. Although an interesting phenomenon in itself, the physical mechanism of matter recycling near the singularity has not been investigated in detail.

After being produced close to singularity, CMB photons stream from the "surface of transparency" (the Ellis et al. equivalent of the surface of last scattering), while losing energy to the gravitational potential well of the singularity, thus exhibiting gravitational redshift, which we observe in our spectroscopes. A consequence of this situation is that the temperature of the CMB is nothing but "antenna temperature" – a parameter of our observations, not something corresponding to the real, physical temperature of space.[55]

As emphasized by the authors of this unorthodoxy, such a model can be considered observationally disproved from the very start. The original paper by Ellis, Maartens, and Nel (1978) showed this model cannot account properly for the (m, z) curve, that is, the relations between the apparent magnitude and redshift of cosmologically distributed sources of radiation. Since this is one of the most basic facts of observational cosmology, there is no doubt that this hypothesis is falsified; as Ellis repeatedly suggested, its point was to better elaborate the notions of singularity and cosmic entropy within the standard picture. For our purposes, we only need to note that the reasons for its rejection are not directly linked to the CMB or its properties. The same applies to a similar model by Phillips (1994a, 1994b). The authors of these two models shared epistemological motivations that led them to build these rather abstract models anew. They both aimed to create a plausible model that would avoid problematic concepts, such as a particular form of singularity marring the orthodoxy, by deploying the full scale of the existing regular fundamental laws of physics, including entropy.

Phillips's model is somewhat more complicated, as it includes two kinds of matter; there are also two singular points, the northern and southern poles (Phillips, 1994a, 1994b). Thus, Phillips called it a "stationary model," in contrast to Hoyle's et al.'s "static model." According to this stationary model, the Milky Way galaxy is located in close proximity to the northern pole of the universe (for the same anthropic reasons as in Ellis et al.'s model). In contrast to Ellis et al.'s model, this one proposes the systematic motion of galaxies from the northern to the southern pole. The motion, however, is laminar and appears stationary, so that the universe, in general, always offers the same picture to a typical observer. Because of this large-scale motion, the observed redshift is partially of gravitational and partially of Doppler origin.

The motivation for the details of the model was twofold: first, to eliminate the unsatisfying notion of singularity in the standard Big Bang cosmology, as we have mentioned, and second, to avoid never-ending revisions of the nature of intergalactic grains that thermalize the radiation with which Hoyle et al.'s and similar models struggled. The thermalizing medium is located near the annihilation region and does not need to fill the intergalactic space (Phillips, 1994a, 775).

It is harder to disprove Phillips' model than Ellis, Maartens, and Nel's model (1978), as the gravitational and Doppler redshifts are delicately entangled in the former. The cleanest test could be the measurement of the peculiar motion of distant sources with respect to the universal reference frame as defined by the microwave background radiation. This measurement is possible in rich galaxy clusters by means of the Sunyaev–Zel'dovich effect (Sunyaev & Zel'dovich, 1980). The prediction of Phillips's model is that more distant clusters will tend to have significantly larger peculiar motions than nearby ones, in direct conflict with the standard theory of structure formation with λCDM. Recent measurements indicate this is not the case, however, and there is no meaningful way to save the theory (Phillips, private communication to one of the authors).

As far as the CMB origin is concerned, these models do not go far from the main idea of the standard model; that is, they associate the CMB origin with the physical state of matter *close to the global singularity*. Only the nature of this singularity is different in the standard lore. Hot plasma near the singularity plays the role of an almost perfect blackbody, and its radiation is redshifted in one way or another to form the observable CMB. While in the standard model, the CMB is created once, and while this creation lies deep in our past (as measured by universal cosmic time), in the models of Ellis et al. and Phillips, the source of the CMB is coexistent with us.

The very existence of these models, not to mention their internal theoretical consistency, demonstrates the falsity of the prejudice often repeated by uncritical supporters of the standard paradigm that only Big Bang theory offers a natural origin of the CMB photons. The CMB cannot be considered in isolation, or only in conjunction with the latest fads and fashions in cosmology (things like baryon acoustic oscillations or primordial gravitational waves or primordial 21cm emission). In fact, the CMB as a physical phenomenon is deeply entangled with the foundational concepts of physical cosmology itself: the notions such as singularities, horizons, and the cosmological principle. Their interrelationships are more numerous and far deeper than the streamlined textbook accounts would have us believe. In addition, our basic dynamical theory in cosmology, namely the theory of general relativity, allows many more solutions if the cosmological principle of homogeneity and isotropy, which is partly a metaphysical conjecture, is abandoned. The role of inhomogeneities in the observed CMB is less significant in these models than in standard cosmology. Their stationary nature puts the issue of structure formation on an entirely different physical footing.

The particular feature of the proximity of the CMB source to the global singularity is shared by another *very* unorthodox model that does contain a Big Bang, but also could, in some readings, be regarded as a kind of steady-state world machine. While it could be classified with the moderate unorthodoxies of Part IV, since it accepts not only Hot Big Bang but also Friedmann models, its deviation from orthodoxy is so sharp that we would be hard pressed to sell it as anything "moderate."

Distinguished physicist, astrobiologist, and popularizer of science Paul C. W. Davies started his long and fruitful scientific career as a cosmologist interested in the cosmological aspects of the arrow of time and the absorber theory of radiation. In a 1972 brief paper in *Nature*, Davies offered a strong contender for the "craziest" alternative hypothesis for the CMB origin that turns the standard interpretation on its head: instead of opaque plasma near the Big Bang, in the CMB, we see opaque plasma near the "Big Crunch," that is, a future singularity in which the closed Friedmann model with $\Omega > 1$ will recollapse. But it is not your garden-variety Big Crunch – it is similar to the Thomas Gold model in which the arrow of time is inverted. 3 K isotropic blackbody radiation is explained in terms of the accumulated starlight from a subsequent time-reversed cycle of the universe, thermalized in the adjoining dense state. Hence, instead of being an "echo of creation," as some astronomers and popular science writers poetically describe it, the microwave background should perhaps be interpreted as a "warning of doom." While the sources are likely to be distributed over space in an inhomogeneous and clumpy manner, it seems clear that the thermalization at the opaque region close to the future Big Crunch will not present a problem, since the large density of cosmological matter, highly ionized by that point, will present a scattering substrate that is as efficient as the zero-mass surface in Hoyle's 1975 model, discussed in Chapter 19.[56]

Davies's paper was written in the recognizable lively style that made him an efficient promoter and popularizer of science; his schematic diagrams showed the re-collapsing model for which the hypothesis could work. It was positively commented upon by Michael G. Albrow of CERN, who believed it offered "a viable and interesting basis for further investigation" (Albrow, 1973). Other than that, it attracted little attention, presumably because of the sharp loss of interest in re-collapsing models after the 1970s. While it played only a small historical role, Davies's model shows how strange possible explanatory ideas could be in the early years of CMB studies. Subsequent precision cosmology cleared the deck quite efficiently: in contrast to Hoyle's hypothesis, Davies's can be falsified by a conceptually simple (although technically quite challenging) direct measurement of the CMB temperature at high redshift (e.g., Molaro et al., 2002)

22

CMB in plasma cosmology

A particular type of unorthodox cosmological model, plasma cosmology, was proposed by Swedish physicist Oscar Klein and developed and defended with great vigor by his compatriot, Nobel-prize winner Hannes Olof Gösta Alfvén (e.g., Alfvén & Mendis, 1977; Alfvén, 1979, 1990). It reached a wide audience with its popular exposition in a book by Eric Lerner (1991), also a plasma physicist. Simply stated, plasma cosmology argues for symmetry between matter and anti-matter, with rather slow annihilation, which could, in principle, provide the energy contained in the CMB (and much else).

Matter-antimatter asymmetry (or baryonic asymmetry) is one of the great mysteries of physical cosmology. It is hypothesized – following the seminal work of Sakharov (1967) – that in the epoch of baryogenesis, in a sufficiently early universe (at temperatures similar to 2×10^{12} K, within the first 0.01 s of cosmic time), non-equilibrium particle reactions coupled with marginal nonconservation of baryon number and a violation of CP symmetry led to a small imbalance of baryons over antibaryons. The imbalance was again a minuscule effect, of the order of 10^{-9}, so the explanandum is clearly real enough, and our theoretical understanding is very incomplete. It was certainly mysterious in Alfvén's lifetime, so there is nothing controversial about his attempts to build an alternative cosmological model on the basis of matter-antimatter symmetry. However, his cosmological efforts have not been widely appreciated or accepted, especially since he accompanied them with sharp criticism of some of cosmology's established methodological practices (e.g., Alfvén, 1984).

Eric Lerner stands out among his few followers. In several papers, mostly published outside the "core" astrophysics magisteria, and in his book, Lerner tried to formulate an alternative account for the CMB properties (Lerner, 1988, 1995). His motivation was both explicit and quite general (Lerner, 1995, 62). He criticized the dark matter hypothesis that is supposed to harmonize the Hot Big Bang model with the measured low density of baryonic matter and the inflationary "paradigm" as *ad hoc* solutions to the problem of noncontact between equilibrated parts of the expanding universe. However, this is mostly a rehashed version of the story of helium thermalizing dust production in early supermassive stars, as discussed earlier.

There is, however, one novelty in Lerner's account – his claim that the intergalactic medium is a strong absorber of radio waves. This absorption is supposed to occur in narrow filaments, with tiny holes scattered about randomly so that distant compact radio sources like QSOs and radio galaxies can be occasionally seen through the holes: "Thus, if our hypothesis is valid, the microwave photons we see were last scattered a few million years ago, not 15 billion years ago" (Lerner, 1988, 464). He (wisely) refrained from discussing how much baryonic matter must be locked in those intergalactic filaments, nor did he dwell on the total real population of QSOs and similar sources that must be excessively large to account for the size of the observed subpopulation; if we see 100 people in a room while peeking through a randomly positioned keyhole, it is reasonable to conclude that the room contains very many people. Lerner did discuss the fit with the *COBE* results in his 1995 paper, however.

The irony is that, in the end, Lerner could not sustain the explanation based on plasma cosmology alone and took refuge in tired-light ideas, eerily similar to Fahr and Zönnchen (2009)![57] This, more than anything else, demonstrates how low the stock of "substantial" Big Bang opponents fell after *COBE*.

23

CMB in non-expanding models

At a first glance, it makes little sense to discuss the problem of the origin of the CMB in non-expanding world models for precisely the same reason why it made little sense to worry about corrosion attacking the steering apparatus and funnels on the *Titanic*. These models had usually sunk many years before CMB was discovered or even speculated about. There is little possibility of any of historically non-expanding models having an impact, from Einstein's original static model, over the many tired-light hypotheses, up to the fractal universes, on the debates related to the spectrum or anisotropies of the background photons. Hence, even relatively well-developed alternatives, such as the plasma cosmology of Alfvén and Lerner, invoke the tired-light idea exclusively at a minor, circumcised point, as discussed in Chapter 22.

Yet non-expanding models started with Einstein, and they usually retain other features of physical cosmology, notably the cosmological principle, so their exploration can be seen as a testing ground for those other more relevant notions. Some might contain ideas that could be repurposed in a manner similar to those of Hoyle and McCrea when their field with negative energy was repurposed in the inflationary scenarios. Admittedly, most of the non-expanding models have little relevant CMB-related content, but there is one interesting twenty-first-century exception.

In a fairly recent work, Wilfred H. Sorrell (2008), then at University of Missouri at St. Louis, made the CMB the key piece of evidence in his non-expanding model of the universe. Instead of an expanding universe, he postulated the cosmic ether, composed of two particles with an opposite charge (aetherons and anti-aetherons). The energy for continuous production of these particles is a result of quantum-gravitational fluctuations, and their interaction releases traveling photons that lose energy over time and are detected as the redshift of light coming from distant galaxies. The main physical assumption of the photon energy loss comes from the work of David F. Crawford (1987a, 1987b, 1991) on the tired-light hypothesis. Sorrell provided details of nucleosynthesis that results in aetherons and anti-aetherons and the mechanism of releasing the photons we detect as the CMB. Aetherons and anti-aetherons are rotating electrically charged dipoles emitting radio and microwave radiation. Thus, both kinds of the particles in the material vacuum continually emit and absorb electromagnetic dipole radiation, resulting in a thermodynamic equilibrium; the electromagnetic dipole radiation has the shape of the blackbody spectrum and a temperature of 2.77 kelvin (Sorrell, 2008, 63). Yet in realistic conditions, Sorrell predicted small anisotropies that are detected by *COBE* and other probes (2008, 69).

The motivation for the model was a strong (perfect) cosmological principle (Sorrell, 2008, 59), postulating isotropic and homogenous universe at all epochs. Sorrell saw the dilemma of the strong versus the weak cosmological principle as the key to future cosmological endeavors (2008, 70). In addition, the blackbody background radiation should depend only on fundamental constants of nature, a motivation similar to that of Hoyle and colleagues. We find it somewhat refreshing to encounter interest in the "fundamentals" of cosmology as a physical science among regular researchers and faculty members. Overall, however, Sorrell's bold model reads much more as a cautionary toy model for the prevailing orthodoxy than as a fully developed framework to replace it.

Part VI

Formation of the orthodoxy and the alternatives:
Epistemological lessons

24

History and epistemology

The emergence of orthodoxy

We have seen several interesting trends in the dissent on the CMB origins. In the incrementally emerging mainstream cosmology, that is, the Hot Big Bang model and a corresponding CMB interpretation, in the first few decades after the discovery (roughly 1965–1989), no single "heresy" came close to attracting wide attention. The most seriously debated unorthodox solution to the CMB origin puzzle was likely Rees's model (1978). This is understandable, as the model contains most of the alternative ideas and mechanisms in a nutshell. As we have seen, there was a small renaissance in the proliferation of alternative models due to a series of measurements at the end of the 1970s and the early 1980s. Even so, the general attitude to alternative ideas steadily and substantially shifted from the mid-1970s to the mid-1990s. The shift became very apparent with the incoming *COBE* results, but the less obvious and more gradual acceptance of large quantities of nonbaryonic dark matter and the relegation of baryonic matter to a secondary role were other contributing factors.

When they did discuss the alternatives, the defenders of the emerging orthodoxy almost uniformly preferred to deal with one "heretical" idea at a time and to avoid quoting similar ideas – clear signs of a careful theoretical debate. There was never any sign of summary dismissal or witch hunts, however, and the debate never spilled into the news media with all the public acrimony that might follow. We could argue, in fact, that there was more bad blood between some of the main supporters of the standard cosmology. One example is the infamous split between the main *COBE* investigators and subsequent Nobel Prize winners John Mather and George Smoot when the latter spilt the information to news media (Mather & Boslough, 1997). What has not been sufficiently appreciated about the episode is the significant media interest in a discovery that was predicted in detail and fully expected within the standard paradigm. This represents a significant shift in attitude, quite unlike the times and style of reporting during the great controversy, when rare media exposures of cosmology tended to focus on juicy tabloid details like the personal relationship of actors such as Hoyle and Ryle.

There was also an absence of consensus among challengers, while the emerging orthodoxy quickly achieved consensus. We can at least partially explain this by the widespread impression that the Big Bang scenario itself is inviolable; it may be slightly modified but not radically rejected, so there is no need for a unified alternative. This did not apply to the indefatigable Hoyle or (at least partially) to the school of younger Indian astrophysicists inspired by him (Narlikar, Wickramasinghe, and Rana). Otherwise, it was often the case that distinguished dissidents, while not exactly accepting the orthodoxy, simply moved to another area of research; this was the case not only with Bondi and Gold, but also with even more renowned personalities such as Fritz Zwicky and Dirac. Insofar as some degree of opposition consensus emerged, it happened in a haphazard fashion and was fixed on the most speculative and *ad hoc* aspect of alternative theories, graphite and metallic whiskers as the thermalizing agents. Nor can we observationally sample interstellar or intergalactic dust for grains of a particular kind (and will be unable to do so for quite a few centuries to come). The consensus looks very much like tailoring the reality to suit the theory. This may be true of the orthodox account as well, but to make matters worse, there was no theoretical consensus that the proposed grains would do the explanatory task even if they really existed, as noted in a careful study by Li (2003).

And the dissention about the origins of the CMB has never truly become a controversy in the sense Kragh ascribed to the steady-state versus Big Bang controversy in the 1950s and early 1960s (Kragh, 1996). Scientific controversies are typically stages in the development of science characterized by unusually prominent political or ethical disagreements, or at least relevant arguments couched in such terms, and entangled with theoretical and factual disputes. They are expected to be resolved by careful appeal to facts and sharpened theoretical principles (Engelhardt & Caplan, 1987). The discussions of the CMB have never contained any unusual social and methodological elements, in sharp contrast to the great cosmological controversy of the 1950s and early 1960s between the classical steady-state theory and the Friedmann models, the latter of which became the standard cosmology. With the exception of theories suggested by Hoyle, Narlikar, and Wickramasinghe that stemmed from much wider unorthodox cosmological schemes of steady-state or "quasi" steady-state ideas, all other unorthodox CMB theories were firmly entrenched in the same milieu as the modern mainstream cosmology outlined at the beginning of the book. Plasma cosmology has never become a serious contender, and from the beginning, closed steady-state models were conceived as thought experiments, not attempts to explain physical reality.

Almost all of the proponents of the hypotheses presented in Part IV fully agreed on a set of major premises: 1) local conservation laws are universally valid, and the underlying dynamical theory is Einstein's general relativity; 2) the universe is, on average, homogeneous and isotropic; 3) the universe started in some form of a Big Bang.[58] This agreement may not be surprising: despite a wide range of diverse views of the relevant phenomena they covered, the sources of most unorthodox CMB theories were observational facts (e.g., the deficit of baryonic matter compared to the nucleosynthetic constraints, or the existence of hierarchical cosmological structure), completely uncontroversial, and fully admitted by mainstream scientists as difficulties, or at least as phenomena in need of explanation.

Discussions of one heretical idea at a time changed to wholesale criticism as new observational data solidified support for the standard paradigm and weakened rival interpretations. A particularly salient example is the effect the arrival of *COBE* data had on proposals on the origin of the CMB. A study by Wright et al. (1994) discussed and rejected a number of unorthodox ideas and models. Some were moderate unorthodoxies, like the high-baryon-mass universe of Gnedin and Ostriker (1992), while others were clearly unorthodox, notably the early thermalization model of Layzer and Hively discussed earlier. In a nod to these models, the authors state, "Many useful models were developed when, in the mid- and late-1980s, it was thought that there could be large deviations from a blackbody" (Wright et al., 1994, 450). The reference here is to the results of the measurements by Woody and Richards and other measurements that corroborated them and their effect on the alternative models. The paper explicitly considers these models' limitations in light of synoptic evidence provided by *COBE* and other sources and discusses the details of the experimental methodology of the *COBE* measurements to dispel any traces of doubt concerning the evidential refutation of the models based on their limitations. Each proposed source of various spectral distortions is listed, discussed, and refuted; alternative models are briefly explained, and their limitations are highlighted considering the results.

Practically all unorthodox CMB explanations were obviated by the set of *COBE* data, and even minor contributions of non-primordial origin became severely limited. By the time of the *WMAP* and *Planck* missions, nobody bothered to refute the remnants of troubled unorthodoxies.[59] In a sense, the *COBE* experiments ended a decentralized and subdued controversy with a convergence on the Hot Big Bang cosmological model and the fireball interpretation of the CMB. In retrospect, we can see all the diversity and varying ramifications of unorthodox theories. The lack of heated public controversy able to match the great cosmological controversy before the *COBE* results masks the fact that the debate was substantial and careful, there were many interesting arguments and counterarguments, as well as rhetorical ripostes, and the formation of the orthodoxy was not an immediately done deal – far from it. This is not surprising, as convincing empirical reasons that could trigger overwhelming consent simply did not exist. In fact, the debate was driven by theoretical insights and preferences that, on their own, could not deal a devastating blow to challengers. Rather, careful theoretical consideration of each of the models prepared the way for the consensus established when the *COBE* evidence concurred with the interpretation regarded as theoretically best motivated. In other words, the alternative models played a decisive role in the formation of the orthodoxy as checks and tests of its weaknesses; only in retrospect can it seem to be an independently developed approach. There is nothing very different from other near-controversies in the recent history of physical science, such as the debate on the origin of gamma-ray bursts in the 1970s and 1980s or the conflict between proponents of the S-matrix theory and adherents of the quark model/quantum chromodynamics in the 1960s.

This theoretical building-up of consensus was partially because the wiggle room for alternative interpretations is much wider in cosmology than in experimental physics; as it is essentially observational science, it provides much more direct evidence to opposing sides in debates (see Chapter 9). And the underdetermination of theoretical accounts by evidence is bound to be much more pronounced and longer lasting in cosmology (see Chapter 8). The CMB was clearly a landmark discovery, but it would be misleading to think it was as effective, say, as the ability of evidence of the existence of an elementary particle delivered by a collider to break a stalemate between competing theoretical approaches. And it would be equally misleading to predicate an historical account on such a view, as we would not see the actual historical trajectory. In general, failing to understand the subtleties of the history of the establishment of orthodoxy runs the risk of eliciting the widespread prejudice that only a few insignificant opinions dissent from the standard paradigm. In our case, even the bibliography of relevant work speaks volumes (literally!) on the unsoundness of this prejudice. Scientists connected with various nonstandard hypotheses included some of the most authoritative figures of twentieth-century astrophysics: Paul A. M. Dirac, Sir Martin Rees, David Layzer, Geoffrey Burbidge, Jeremy Ostriker, and Sir Fred Hoyle.

The prolonged episode of explaining the CMB in the history of modern cosmology wherein underdetermination resulted in varied alternatives to the emerging orthodox view of the CMB goes against both sides in the debate on the underdetermination of scientific theories as Stanford (2006) has characterized them. On one side of the debate are those defending the view that if alternatives to the orthodox view can be easily produced within the scope of the existing evidence in science, then underdetermination is an acute state of science. These are typically proponents of the logical version of the underdetermination thesis we discussed in Part III; they try to come up with procedures and algorithms to produce alternatives for any scientific theory. On the other side are defenders of the view that any attempt to come up with viable alternatives faces "profound difficulties and rare success" (Stanford, 2006, 16), so there are virtually no alternative theories in scientific practice. Thus, underdetermination may be only a fleeting state in a few cases. In our episode of interest, the interpretation of the CMB, the richness and acute relevance of all the alternatives to the orthodox view and their role in eventually establishing its supremacy indicates that, at least in cosmology, if not in other historical natural sciences dealing with the deep past, viable alternatives are neither nonexistent nor necessarily rare, and their presence does not necessarily imply an acute hopeless state of underdetermination in the field. One crucial reason why both sides in the underdetermination debate drew wrong conclusions may be that there were very few well-documented scientific efforts to produce alternatives before the period when our episode unfolded. The moral of the story for philosophical explorations of underdetermination could be that they should be supplanted and informed by detailed case studies of potentially relevant post-WW2 episodes across sciences. In any case, the prolonged state of underdetermination in cosmology was not an impediment but a necessary stage of critical assessment and convergence.

Overall, there was one prior step in response to both the proponents of the logical version of underdetermination thesis and the deniers of the thesis that preceded construction of the unconceived alternatives similar to those Stanford (2006) pursued. Before asking what was not conceived, we can consider what was conceived in episodes similar to the one we discussed. This goes to the heart of scientific practice. Once we get a clear understanding of what production of the alternatives meant in a crucial historical episode, this may boost the possibility of producing unconsidered viable alternatives in contemporary cosmology and other methodologically related scientific fields. The recurrence of acute underdetermination in a field (Sklar, 1975) may not be a reason to embrace anti-realism as Stanford (2006, 17) suggested, but rather a sign of the epistemological health of a developing field climbing to yet another level of understanding and precision.

It has also become clear that setting up an unqualified conceptual dichotomy between "facts" and "theoretical claims" could impede historical analysis. We discussed Bondi's argument (Chapter 8) developed in the 1950s and aimed at the erroneous and damaging use of such dichotomies in scientific practice. The simplistic view behind such a hasty and insufficiently qualified dichotomy was (and still is) a legacy of the era of the dominance of logical-positivist views of scientific theories and scientific knowledge in general. The logical-positivist focus on understanding theories and hypotheses as axiomatic (formal) structures consisting of theoretical sentences that are then tested by *independent* observational sentences (and theory is refuted or confirmed in light of the test) does not necessarily require a more complicated picture. The view is certainly correct at a very abstract level, and it may have its uses in philosophical discussions devoid of the actual scientific content. It has been invoked as a response to various sorts of relativistic views of scientific knowledge. But it tells us very little, if anything, about the subtleties of real scientific practice. Thus, forcing such an abstract view onto a case analysis impedes it. The notion of fact is simplistic, if not qualified, in terms of the observational and experimental evidence. And as Bondi stated, in actual scientific practice, observational statements can be less reliable than theoretical ones.

What the episode we discussed demonstrates is that to arrive at wide convergence on both "facts" and theoretical explanations, we need a wide array of various and often conflicting theories, models, and observational resources, each with its own limitations. A host of features of the orthodox cosmological model – with each individual feature supported by distinct lines of observations of varied certainty – had to be accepted in order to accept the emerging orthodox view of the CMB. Concomitantly, it must be clear that each of these features could be questioned by the next balloon, rocket, or satellite experiment. The same was true for each alternative view. A detailed analysis of actual scientific episodes often makes short work of hasty philosophical dichotomies that are produced in order to argue at a very different level of understanding scientific knowledge. Thus, even if the acceptance of a scientific theory is justified by facts, generally speaking, when we look at the process that leads to the acceptance of a theory in a scientific field where underdetermination is pronounced, we may find that prolonged interaction or rather a continuous trade-off between available evidence and theoretical views may be crucial but may only gradually determine a complex domain of facts and observationally justified theoretical claims. An inadequately qualified approach to the facts/theory distinction can lead us astray in two ways. First, we may overlook alternatives and their importance in philosophically motivated analysis of historical episodes. Second, we may unjustifiably draw firm lines demarcating firm factual domains.

The same may apply to other scientific fields, but it becomes more obvious in fields like cosmology. This brings us back to the comparison of scientific disciplines that can substantially intervene in the phenomena they are examining via experiments and those that are bound to use more passive observational techniques (Chapter 9). The spaces for relevant theoretical views on the one hand, and the observational domain on the other, are interdependent. Thus, the uncertainty of evaluating the temperature of the CMB was tied to the atmospheric distortions across its layers and the ability to perform measurements in the vacuum in order to minimize the thermal properties of the surroundings. Eliminating atmospheric distortions and only dealing with the "quality" of the vacuum eliminated a host of alternative explanations, as some previous deviations from the shape of the blackbody spectrum could be attributed to atmospheric distortions. In experimental physics, the use of lab temperature measurements avoids the impediments of the upper atmospheric layers, a major source of confounding in the CMB measurements, thus immediately narrowing the field of alternative ways of connecting the key parameters.

Notice, moreover, that *by the same token*, this "interventionist" view limits itself to a local environment in terms of space and time. Just as the elimination of atmospheric distortions led to the elimination of alternative theoretical narratives, we could hope that repeating the same procedure on a higher level, namely eliminating the distortions due to the Milky Way emission and absorption, would lead to further theoretical refinement. The hope would be naïve, unfortunately. We cannot launch a telescope into intergalactic space beyond the confines of the Milky Way and are unlikely to be able to do so in next million years or more. While we can easily perceive some problems that could be overcome, in principle, by removing the observational obstacle (especially those with low-ℓ moments in the multipole expansion of the CMB map to be discussed in Chapter 28), doing so is simply unrealistic.

When the domain of relevant observational resources is more or less reduced to satellites that bypass existing observational uncertainties, the domain of relevant theoretical options shrinks too. A reliance on heavily experimental fields that deal with smaller domains both observationally and theoretically may lead us to hastily accept a simplistic conclusion about the fact/theory dynamics. The actual evidence does not provide immediate theoretical convergence because evidence is typically a rich domain and rarely a simple fact. Nora Boyd (2018) recently made a very important point about "enriched evidence" in science, noting all the twists and turns of specific contexts (theoretical and observational) and limitations across various examples, including astronomy. The process of discovery within such contexts shapes the domain of "facts" and theoretical claims to which the community converges and strengthens the "justification" of theoretical claims. As we have pointed out following Ellis (2014), physical limitations in the context of cosmology are unique; the object of study, the emergence of the universe, is a unique event, the observational horizon is limited to 40 billion light years, and light-carrying observational signals travel at a physically limited speed. These physical limits, combined with the observational challenges of the CMB measurements, were overcome gradually over the course of several decades, providing a slowly developing mutual feedback loop between theories and observations that eventually led to the overwhelming convergence.

25

What about the alternatives?

The story of the CMB origin offers insights into the nature of the progress of modern science – its good and bad points alike. The role of the empirical but unexpected discovery of the CMB as unraveling the deepest mysteries of the origin of the universe was immediately and widely recognized by almost the entire cosmological community, including most researchers with unorthodox views. And generally speaking, it helped persuade a large portion of the wider scientific community that cosmology is a serious, mature, and firmly founded scientific discipline.[60] The cutting of the Gordian knot of the great cosmological controversy opened up new vistas in cosmology as well. The N-body simulations of galaxy formation, investigations of the power spectrum of density fluctuations (López-Corredoira, 2013) or the correlation and autocorrelation functions of various collapsed structures, the search for primordial particle relics, or theories of the QSO absorption line systems, to mention only a few, comprise the most sophisticated layer of cosmological analysis to date predicated on the discovery of the CMB and the effort invested in its standardized interpretation.

The number of studies devoted to the standard interpretation of the CMB is overwhelming. This is not necessarily a good indication of the number of alternative accounts, however, as most studies explicitly assume the standard interpretation is valid, including those that reconsider various aspects of it. Thus, we find much serious reconsideration of the key aspects of the orthodoxy – for example, the relations between the CMB anisotropies on the one hand, and the quantity and the kind of structure in the standard λCDM model on the other[61] – sheltering under the umbrella of the paradigm's general acceptance. Similarly, serious considerations of the possibility that General Relativity may not apply cosmologically if the CMB is due to discrete sources appear in the work of key proponents of the orthodoxy. The statement of unqualified acceptance is often a rhetorical device used to prevent hasty negative conclusions on the status of the orthodoxy by either professionals or the wider public. This is the aspect of knowledge production in cosmology that sociological studies of the discipline ought to take into account. The situation may be a specific and not necessarily desirable form of "paradigm defense" in the Thomas Kuhn's sense (Kuhn, 1962) used by mainstream cosmology. Given all this, it is understandable but perhaps regrettable that the number of textbooks in which alternative interpretations are even mentioned is insignificant; counterexamples are usually found in textbooks written by the "mavericks" themselves (e.g., there is an interesting and open-minded account in Narlikar, 1983).

Overall, dissent has served the lofty principles and ideals of scientific enterprise.[62] As we have seen, given the epistemological and methodological ramifications of modern cosmology it would be rather awkward if alternatives had not flourished – as Ellis (2014) pointed out, one does not fully understand a model without understanding its alternatives. Alternative explanations of the CMB origin have, in most cases, been falsified, but this has led to a new problem-situation permitting the emergence of new views, such as inflation or quantum cosmology, since the 1980s. It should be clear at this point that the dramatic transition from the original problem-situation set up by the steady-state challenge to the relatively poorly defined relativistic orthodoxy in cosmology to a completely new level of high-precision cosmology and studies of the very early universe was occasioned not just by a single momentous, epoch-making empirical discovery, but also by a blizzard of theoretical activity surrounding it, including many iterations of conjectures and refutations, some of which we have noted here. This happened in four stages (see Chapter 6), so that incrementally, over the course of three decades, the alternatives' maneuvering space was reduced.[63] Even so, the alternative explanations have an ongoing role to play. We will draw three methodological lessons from their history, and then look at some of the ideas they harbored, as they may still be relevant.

First, most alternative models accounting for the CMB have never been fully developed and are not even close to the level of detail of the Hot Big Bang model. As we have suggested, this may have to do with the lack of consensus among the unorthodoxies – they just did not have enough common ground to pull the resources together. In addition, some alternatives were toy-model reactions to the model that was becoming dominant. Yet our analysis indicates that the level to which a researcher can improve an alternative model overall should not be underestimated. For example, Reese's second model was a thorough refurbishing of Layzer and Hively's appealing yet deficient model. In addition, the account of the CMB within the steady-state model and its better fit until the *COBE* data arrived demonstrates how far a comprehensive and elegant but unorthodox model can go in accounting for the main physical facts. And even an incorrect model can be helpful in identifying the weaknesses of unorthodoxy, as was the case with the isotropy of the CMB in early versions of the standard model.

Second, it is possible to develop alternatives employing a piecemeal rather than a wholesale approach. One of the lessons of our analysis of the history is that a general cosmological framework can be clearly distinguished in the explanations of the CMB origins, and the latter can be considered and developed independently. Thus, there may be a number of routes we can take in rethinking the details of physical significance, only subsequently turning to a more general framework. In fact, in a more mature phase of cosmology when observations abound, these two points become increasingly salient, as we are able to make much more abstract assessments. Thus, for instance, we could turn the tables and ask what sort of desiderata an adequate model would require to fit the existing precise observations of the number of free parameters in the broadest possible sense. The orthodox solution may simply seem to be one in a set of possibilities, "better than would be expected if it were a wrong theory" (López-Corredoira, 2013, 1350032–15), but only somewhat more likely than a host of other fits. For example, López-Corredoira (2013) analyzed density fluctuations (acoustic peaks) in the power spectrum of the CMB in this manner, but the approach could be extended in a comprehensive computational project that would generate plausible alternatives capturing relevant free parameters across the domain of available observations.

Third, we should bear in mind that there might be less apparent theoretical presuppositions lurking in the background of the orthodoxy, and these may motivate and make valuable certain alternative explanations of physical phenomena, such as the CMB. Devising models and explanations of key physical phenomena rests on the intricate interplay of theoretical presuppositions and selected observations. A particularly instructive case is the requirement for large quantities of baryonic dark matter in more recent cosmology; this fits well with the earlier postulation of thermalizing grains to explain the CMB as non-primordial in Layzer and Hively's model. Similarly, with new evidence of particular facts, for example, those pertaining to baryonic matter, alternatives that seemed unappealing suddenly become plausible. This is certainly not to be understood as a claim that the Layzer–Hively hypothesis has again become viable. Instead, it is a claim that the value of individual building blocks in the entire explanatory construction can and does vary with time; in this instance, a previously discarded block of the Layzer–Hively structure has incidentally been found more solid than hitherto thought, although the structure as such remains nonviable.

For all these reasons, it is instructive to nurture a scientific community that actively develops alternatives to the orthodoxy and allows bold conjectures and fringe models.

Another question to consider is what sort of an edge, if any, the mainstream interpretation of the CMB and the inflationary approach in general has over alternatives, given the widely accepted criteria of what constitutes scientific evidence. Astroparticle physics can provide some evidence of particle properties, as for instance, cosmological constraints on the number of flavors and masses of neutrinos (Steigman & Strittmatter, 1971). But the cosmological evidence is, on the whole, very different from the evidence provided in, say, experiments in solid-state physics, as we demonstrated in Chapter 9. And in the case of particle physics, the crucial aspects of theoretical models are much more directly tested in particle colliders. Thus, we can never test the primordial fireball hypothesis directly, as we can myriad other physical parameters in a laboratory. To clarify the nature of the evidence upon which cosmology is based, researchers sometimes compare it to the sort of constructive evidence found in archaeology or paleontology. We may well ask to what extent this sort of analogy is adequate. We may even doubt that cosmology meets these standards of evidential support because it is so indirectly related to the core of theoretical models and leaves them open to underdetermination to such an extent that it is questionable whether we can label the evidence supportive, instead of merely indicative.

The reference to paleontology is apt, as we argued throughout Part III, at least in terms of the theory–evidence relationship, but cosmology may be even more limited due to the physical limitations we pointed out. We should note, however, that cosmological evidence has some important historical advantages, such as much earlier attention to the so-called selection effects that can be detrimental to constructive evidence. Yet on the whole, we do not think the viability of a scientific field will necessarily be questioned simply because it does not meet the stringent standards of evidence set by experimental physics (and probably only in some areas of it). In terms of the standards of evidence, however, the distance between experimental physics and cosmology in general, including the CMB case and its interpretations is much greater than that between various cosmological approaches (including astroparticle physics). The latter are all confined to a common standard of evidence that is substantially different than the standard applied in most particle physics.

Now, theoretical accounts based on such kind of evidence may be prone to more or less reticent underdetermination as we discussed in Chapters 8 and 24, even if the rivals are treated as falsified with a great deal of certainty. Given this, it may be useful to avoid treating failed alternative interpretations as we would treat falsified alternatives in experimental physics, that is, as theories to be straightforwardly discarded. Instead, it may be more advantageous to regard them as a resource for approaches that can potentially but realistically be revised and revived or as initial dips into a wider pool of possible alternatives. Even though the plausibility of the existing alternatives to the orthodox explanation of the CMB has diminished with advances in detection techniques, the nature of the evidence leaves the field much more open to various revisions of the existing alternatives or to the development of their various elements than, for example, particle physics, where models and theories have been straightforwardly discarded considering the experimental results.

Following this rationale, it may be useful to rank the unorthodox CMB explanations in terms of the plausibility and persuasiveness of their theoretical grasp and promise. Such a ranking may provide an overview of the alternatives and suggest a framework for understanding and judging them, perhaps even for developing some of their ideas.

Within the "radical" group, arguably the most interesting solution is the use of the framework of the conformally invariant theory with variable masses (Hoyle, 1975). Its solution for the puzzle of the CMB origin – divergent scattering of photons when passing through the boundary of cosmological domains with different mass signs – is devilishly ingenious. It promises interesting physical insights into such crucial questions as the nature of mass or the epistemological possibility of equivalent descriptions of a single and only indirectly accessible physical event. There certainly are not many cases where the famous Italian saying of *se non 'e vero, 'e ben trovato*[64] is more appropriate. On the one hand, it is a pity that the brilliance and authority of Sir Fred Hoyle have not motivated the development of his ideas.[65] On the other hand, the same cannot be said for the radical unorthodox static models of Ellis et al. and Phillips that are, in a final analysis, currently little more than curiosities, abundant in the history of any sufficiently rich and dynamic scientific field.

Among the "moderates," the idea of discrete sources creating the CMB has been easiest to discard, since it essentially depends on the status of experimental techniques. The latter have seen almost explosive development throughout the last several decades, especially since the discovery of the CMB firmly established cosmology as a respectable user of cutting-edge observational equipment with ever-increasing resolution. Therefore, we are dealing with a sort of bootstrap, so often encountered in young sciences: a bold hypothesis is given emotional support or preference by a majority of the scientific community in spite of a lack of direct support, motivating tremendous observational and experimental efforts whose results solidify the support for the hypothesis, turning it into a paradigm.

Those unorthodoxies involving new physical elements (such as variants studied by Carr, dealing with Population III stars and early black hole properties) are more difficult to deal with, as shown by some of the examples we discussed, but discussions of them remain fruitful to this day. This is true even if we consider them as somewhat beyond the cosmological mainstream. As we have shown here, the mainstream focusing is a highly complex and nonlinear process, with many studies "gathering dust on shelves" subsequently becoming incorporated into the mainstream. Stock examples include Yang–Mills theories, baryon number nonconservation in particle physics (of which Hoyle certainly was a precursor), or horizontal gene transfer in evolution/molecular biology.

Finally, an intricate and important level of skepticism keeps the alternative models alive. In Chapter 8, we discussed Bondi's point that observational insights are always in danger of being less reliable than theoretical ones. Given the physical and temporal depth at which cosmological models aim, there is always a reticent fear that in cosmology, to quote Swarz and coauthors, "the more fundamental the assumption, the harder it appears to test" (Swarz et al., 2006, 184001-1). Thus, the individual harmonics we cannot resolve adequately may always conspire to show us a coherent, but ultimately wrong picture. In fact, these authors explored recent unexpected anomalies in the CMB anisotropies measured by *Planck* and *WMAP* at large angular scales and compared them to the inflationary model's prediction of statistical isotropy and scale invariance of inflationary perturbations. The results of the measurements at large angular scales with two or multiple point functions have been criticized as possible statistical artifacts (as "aposteriori statistic") but the issue is certainly still open. Another anomaly of hemispherical asymmetry, the power of radiation being stronger at one hemisphere than the other and a cold spot in the *WMAP* (Swarz et al., 2006), as we explain in Chapter 28, cannot be resolved without precisely the sort of thinking that drove the physicists who came up with the alternatives post-1965 – the search for more ordinary explanations like foregrounds in the "masking" of observations by phenomena in the solar system or even intergalactic dust grains. Thus, the methodological moves that built the alternatives are now deployed by the inflationary Hot Big Bang model defenders, another indication of how valuable methodologically and educationally the failed alternatives may be.

26

Pragmatic aspects of model building and social epistemology of cosmology

The scientific explanations in the CMB episode in the recent history of modern cosmology we discussed target a rather narrow physical phenomenon. Relevant observations basically boil down to a series of focused antenna measurements and the ensuing exploration of the signals' uniformity. Despite this narrow focus, an extremely rich variety of models have been employed to explain the phenomenon. These models are so diverse that they pretty much cover all the basic scientific models identified and categorized by philosophers of science (Frigg & Hartmann, 2018)

Some explanations stemmed from various cosmological models: the steady-state and its modifications, the Cold and Tepid Big Bang models (Chapter 11), the closed steady-state (Chapter 21) and non-expanding models (Chapter 23). The observed properties and predictions were inferred in accordance with these models. Some were constructed as elaborate or toy explanatory models of the CMB, rather than one-shot explanations independent of a particular model. The primordial chaos explanations (Chapter 14) stemming from the Cold Big Bang cosmological model were probing toy models, to use the label Frigg and Hartmann (2018) put on that sort of scientific model. Moreover, the fundamental theories employed by cosmological models varied (Newtonian mechanics, special theory of relativity and general theory of relativity, but also Milne's kinematic relativity in Rowan-Robinson's hypothesis). Other models, such as Rana's (Chapter 16) or those postulating unresolved sources of CMB photons (Chapter 12), were purposefully dependent only on regular well-confirmed physical laws and theories, circumventing or explicitly shunning any cosmological models or even the prevalent cosmological assumptions.

Thus, there were two kinds of model-based explanations: 1) the subsidiaries to cosmological models and the fundamental physical theories underlying them, and 2) the minimalist models based on regular physical laws. These two categories were motivated by different epistemic approaches and values, and usually such motivations were spelled out. The proponents of the subsidiary models valued explanations that were part of a comprehensive physical-cosmological model. This attitude was more or less expected given that the predictions of the CMB came from the proponents of such a model. Yet this was not a convincing reason for the proponents of minimalist models, who were, generally speaking, skeptical of using anything other than the regular well-confirmed physical laws to explain observations. The proponents of the steady-state theory produced explanations of the first type but with the motivation of using only regular physical laws and avoiding the murky relationship between physical laws and initial conditions. We return to this issue in Chapter 30; at this point, it suffices to say that there were deep epistemic, ontological, and methodological assumptions behind these choices that could hardly be avoided and were often spelled out. Thus, both the approach to the evidence – what exactly the signal is supposed to explain and where it fits in the physical world – and the assessments of the evidence – whether and to what extent it is a real signal rather than an artifact of atmospheric interference or some astrophysical phenomenon – were epistemically motivated.

Without this genuine variety of epistemological frameworks and motivations, the space of plausible alternatives would not have been explored, nor would deficiencies of the orthodoxy have been diagnosed as early. Consequently, observational programs would not have addressed these legitimate worries. This is why both epistemic diversity and empirical diversity – often neglected in historical and philosophical studies focusing on the accounts to which the community eventually converged – are necessary in scientific fields such as cosmology, as the space for alternatives always lurks in the background. The CMB episode we discussed, with all its intricacies and details, is a model of epistemic responsibility in the field in this respect. But how is this diversity achieved, and what sort of structure of the scientific community and funding ensures it?

As discussed in Chapters 2 and 8, Hoyle (Hoyle in Gregory, 2005) and Halton Arp (1987) pointed out two problems with the structure of scientific endeavor in modern cosmology that was emerging at the time. First, increasingly expensive equipment was required to test any of the predictions, and such equipment and its use-time were centralized in a few prominent institutions. Second, theoretical options outside the mainstream became harder to push because of the ever-increasing mainstream production at a few prominent institutions. Now we know that these two problems are much more acute in other areas of physics; for example, studying particle physics at high energies requires vast particle colliders. The challenges the centralization of this sort presents for the scientific pursuit and the ways of handling them have been the topics of numerous studies.[66] Various simulations (Šešelja, 2022; Zollman, 2007), formal studies (Page, 2011), and relevant empirical studies (e.g., Abbasi et al., 2011; Milojević, 2014; Perović et al., 2016) have shown there is a threshold at which centralization starts to increasingly impede the efficacy of problem-solving in scientific institutions. Physicists themselves are acutely aware of the phenomenon that Arp and Hoyle pointed out, and debates on strategies have been going on for a while.

Astrophysics has become big science by any measure, including financial and organizational. This has led, among other things, to the emergence of "telescope time allocation committee bias," as Sarah Gallagher and Chris Smeenk (2023, 216) labeled it. But the extent of its centralization is not egregious, inevitable, or perhaps even possible compared to the extent of centralization in high energy physics. Moreover, astrophysics substantially benefits from a score of other disciplines, including high energy physics, thus inherently decreasing the scale of centralization.

As we have pointed out, cosmology entered the big science club with the dedicated experimental testing of the CMB properties performed by equipment mounted on the $U-2$ aircraft. This trend continued with a string of dedicated satellite missions, such as *COBE*, *WMAP*, and *Planck*. Yet as we have seen, the tendency toward the centralization of the equipment and observational work does not necessarily lead to the diminishing of alternative theoretical accounts. Occasional rocket and balloon measurements of the CMB properties performed in the late 1960s, 1970s, and early 1980s were sufficient grounds for the construction of various alternative explanations of the CMB. Thin indecisive evidence can result in a rich theoretical domain, especially in fields with a pronounced state of underdetermination of theories by evidence.

The other challenge that Hoyle pointed out, the diminishing chances of developing alternatives to the orthodoxy, may still arise in the context of the centralization of theoretical work due to institutional inertia and complacency. The field of theoretical physics has been accused of having such an impediment in relation to the overwhelming dominance of the string theory, by insiders (Hossenfelder, 2018; Smolin, 2006) and relative outsiders (Woit, 2011) alike. We cannot discuss here whether such criticism is justified, or to what extent, but the case of fundamental physics has specific features that can be instructive for cosmology's future. First, it is a mature field, backed by extremely well-tested fundamental theories (quantum electrodynamics and quantum field theory), and all the centralized experimental work concerns various models (most prominently the Standard Model and the Supersymmetry) that accord with those theories. Second, string theory and its alternatives are highly speculative theories detached from experimental work in terms of predictions and testing.

Throughout the history of modern cosmology, alternative models have often specifically aimed at producing feasible tests to attract the attention and the necessary level of confidence of the community. This may contain the recipe for cosmology to stay vibrant in the age of big science and avoid some of the challenges that seem to plague contemporary fundamental physics, more specifically that the construction of new models and alternatives should aim for feasible tests on the existing or anticipated equipment, while more abstract theories and models (e.g., the multiverse theories) should be cooked but on the back burner. Thus, theoretical work should favor testable alternatives, and observational programs should take them into account. Of course, this only holds if there is no strong institutional pressure to exclusively concentrate on the development of the details of the orthodoxy. But even here, there is a lesson in the early CMB debates: it was epistemically responsible for even the most prominent champions of the orthodoxy to allocate some of their time to developing feasible alternatives.

27

Large-scale numerical simulations in cosmology

Beyond the theory–observations distinction?

Conceptually, it is not easy to draw a line between simulations, observations, and theory (Winsberg, 2019). This is especially the case for modern astronomy, astrophysics, and cosmology, as the three are tightly interrelated in scientific practice. Simulations are often a crucial addition to observations; without them, observations would have little value, might succumb to various biases, or may not be meaningfully related to existing theoretical models. Gallagher and Smeenk summarized this situation, stating that in astrophysics, "[c]omputer simulations now play an essential role, because the layers of selection effects have become too complex to account for with simple numerical corrections" (Gallagher & Smeenk, 2023, 213). Even counting and cataloging sources such as stars, galaxies, galactic clusters, and QSOs could not be pursued to the extent it is now without simulations correcting for biases, such as those concerning luminosity or distribution, by incorporating various theoretically plausible assumptions or even anticipating and modeling theoretically plausible sources that may not be observed.

All this becomes crucial, for example, when cosmologists try to square the large-scale isotropy and homogeneity of the universe (the values promoted to the status of a fundamental theoretical value in cosmology by the discovery and exploration of the CMB) on the one hand, with the exploration of expected local departures from such fundamental properties in local structures on the other. Probing such departures requires variations of scales at which the distribution of structures is explored and expected not to be uniform; local observations alone cannot provide these. Moreover, precise probing of regular anisotropies such as the Dipole anisotropy or the probing of alleged anomalies such as the "fingers of God" or "axis of evil" (Chapter 28) requires simulations that test the existence and nature of the anomalies under variations of data-masks blocking the galactic foreground "contaminations" of the CMB signal (Rassat et al., 2014).

It is only natural that such difficult computational tasks are outsourced to supercomputers. In itself, it is nothing new. If anything, some of the protagonists of this book were among the first to notice the need for computational cosmology. The ubiquitous Hoyle was one of the first to insist that in modern astrophysics and cosmology, it is impossible to do well without massive computational resources. Two major biographies have emphasized his focus on this aspect in his organizational activities (e.g., Gregory, 2005, 184, 212–213; Mitton, 2005, 174–175, 294). His cherished Institute of Theoretical Astronomy (IoTA) was equipped with state-of-the-art electronic computers at a time when this was an oddity in scientific institutions. In fact, part of the friction Hoyle encountered at Cambridge was due to the failure of many of his colleagues, even within the mathematics and physics community, to appreciate this point. He was certainly ahead of his time in this respect. A somewhat similar attitude was shared by Yakov Zel'dovich in the Soviet Union.

As in other scientific contexts, the advent of massive parallel numerical simulations transformed the world of physical cosmology. In particular, this applies to the most important theoretical problem of cosmology after the solidification of the orthodox Big Bang view – the origin and evolution of cosmic structure, on scales ranging from individual galaxies ($\sim 10^3 M_\odot$ for the so-called ultrafaint dwarf galaxies like SDSS J1049 + 5103[67]) to the "Great Walls" ($\sim 10^{18} M_\odot$ for all superclusters comprising the Sloan Great Wall[68]). To encompass such a tremendous range of sizes and masses, the structure formation mechanism has to be truly universal and extraordinarily efficient. Since the structures to be explained cover a huge interval in redshift – and hence time, on both orthodox and a large majority of alternative cosmological models – between $z = 0$ and $z \simeq 10$, and sample a huge variety of environmental conditions, it is exceedingly difficult to discern the universal properties of the evolutionary processes involved and to separate them from historical contingency and happenstance.

Yet cosmological simulations perform another crucial function that is rarely explicated, although it is of great methodological and philosophical significance. They offer a glimpse of a *truly Copernican view* of the world on the largest possible spatial and temporal scales. The fact that we cannot pick and choose our position in space (and are severely constrained in our possible locations in time, as per anthropic reasoning; see Dicke, 1961; Ćirković & Balbi, 2020) restricts the empirical database available to any cosmologist in practice, if not in principle.[69] It is not an epistemic luxury: we *have to* reason as if we are typically located observers. Nothing demonstrates that better than the CMB map (Figure 5.3). We have one such map, and while our angular and temperature resolution may improve in the future, there is no way for us to know what the CMB map looks like for an observer located on a habitable planet in, say, the Bullet Cluster (1E 0657-56) at the co-moving distance of about 1.1 GPC. The problem of cosmic variance, with all the fuss related to the low CMB multipoles, belongs here.

Creating synthetic universes, so to speak, represents the only serious way to try to overcome this problem. Modern large-scale numerical simulations like the Millennium simulation (Springel et al., 2005) or the Illustris Project (Vogelsberger et al., 2014) rely on the gravitational dynamics of N bodies plus hydrodynamics and assorted other physics to reproduce a wide range of observable properties of galaxies and the relationships between these properties. Of special interest are properties such as the ratio of the amount of stellar mass to dark matter mass in various galaxies (and its evolution through cosmic time) and the total amount and distribution of star formation in the universe as a function of time. This could have been achieved only with the help of Moore's Law; for example, the largest and most extensive version of the Illustris Project simulation was run on 8,192 processing cores and took *19 million CPU hours*.

Generating simulations inevitably creates selection effects of its own, however, as parameters for simulations are chosen, and trade-offs are made between available computing time and the selection of regions of the entire simulated domain. One potential danger is what Gallagher and Smeenk (2023) labeled the "problem of uncomputed alternatives," analogous to Stanford's (2006) worry about unconceived alternatives to major scientific theories. Simulations inevitably consider "a limited space of alternative proposals" (Gallagher & Smeenk, 2023, 207). Certain fundamental objects, features, and entire scenarios may be physically plausible but computationally intractable, preventing such alternatives from tallying with the observations. In astrophysical simulations, this is often related to the adequate choice of initial conditions and the inability to achieve tractable simulations of a small number of object interactions with appropriate variations of initial conditions (as opposed to the simulation of a large number of such objects with a narrow set of initial conditions) (Gallagher & Smeenk, 2023, 218). A prohibitively sized space of possible parameter values prevents the tractability of scores of potentially plausible hypotheses. This, in turn, determines what sort of observational questions are addressed by actual observational programs.

This worry is certainly acute in cosmology. Yet as we have seen, some alternatives to the orthodox explanation of the CMB opted for very abstract cosmological properties while others opted for explanations based on mundane physics. Perhaps the heuristic rule that can address the worry of missed alternatives due to computational intractability is the simultaneous exploration in various directions to at least some extent, and the diversification of resources to support such an approach. Currently intractable problems due to prohibitive costs can be pursued once a satisfying result has been achieved in a tractable yet potentially theoretically deficient mode. A model for addressing such worry already exists in high energy physics, where the genuinely exploratory combing of the existing vast data on particle collisions was initiated after the major achievement of particle discovery and measurements. The physicists were aware that unanticipated features and properties could be hiding in the big data and searched for unexpected signals. Key particle properties have been discovered in such a way before, after all (Perović, 2011).

Part VII

Other philosophically relevant aspects of the CMB

28

CMB and Copernicanism

"The Axis of Evil" and "The Fingers of God"

Given the prevailing tendency to embrace the Copernican principle stating we are not populating observationally a special place in the universe (except for the proponents of some closed-state models), it would be unsettling to find even faintly convincing observational evidence (as opposed to an interpretive cosmological argument) to the contrary. In their 1992 piece in *Physics Letters A*, Arp and Van Flandern offered a comprehensive case of the "fingers of God" distortion of the redshifts due to expansion. In any direction we look, clusters of galaxies seem to be grouped radially; that is, they are distributed as fingers of elongated concentration of galaxies pointing to the observer, that is, to us. This would certainly be dismissed as a mere accident if it occurred in one cluster here and there. However, it is observed in all directions, motivating some cosmologists to see it as a phenomenon that the Big Bang model cannot explain plausibly. The time needed to realize this sort of non-accidental distribution dynamically is five times longer than the period since the Big Bang. The missing mass that would harmonize such a distribution with the Big Bang timelines is explained by the dark matter hypothesis that also does other explanatory work, most notably explaining the discrepancy between visible and dynamic calculations of galaxies and galactic clusters (Arp & Van Flandern, 1992, 263). Arp and Van Flandern thought the introduction of the missing mass was an unsatisfying move to save the model and argued that redshifts are not the result of velocities of clusters due to expansion but caused by other factors (e.g., Hoyle's mini bangs). Of particular interest is their detailed, five-point observational argument against the velocity-redshift hypothesis central to the Big Bang model. It boils down to an argument for the physical association of quasars and galaxies in the photographs of relevant objects, thus countering the Big Bang hypothesis that they are substantially distant in time and space (i.e., quasars originated much earlier) based on the redshifts.

Now, this indirectly but inescapably knocks out the orthodox interpretation of the CMB. Thus, led primarily by the "fingers of God" argument, but also affected by his dissatisfaction with the nucleosynthesis account of the orthodox view, Arp (1992) presented a very general motivation for many radical and moderate models: irrespective of the current resolution power of telescopes at any given point, he said, it is plausible to suppose that the smooth CMB is a result of unresolved discrete sources: "the long line of sight through the universe will average out fluctuations and make it much easier to account for the exceedingly smooth CBR which is currently observed" (Arp, 1992, 272).

A similar anomaly noticed in 2005 (Magueijo & Land, 2006) is the "axis of evil" (AoE) anomaly where alternating hot and cold spots seem to conspire in an axis across the CMB. This perceived anomaly has been a subject of some debate, discussed as possible evidence of alternatives to the standard model (Schild & Gibson, 2008), and various suggestions have been floated in a rather laconic manner, including a rupture in the early structure due either to unknown quantum fluctuations or to an early contact with another universe. The prevailing view is that it is either a statistical fluke or an artifact masking galactic foregrounds in a particular way (Rassat et al., 2014). What is clear, however, is that the anomaly is not an artifact of the detectors, as both *WMAP* and *Planck* confirmed this.

As these two examples demonstrate, evidence can be considered widely acceptable only if it is enriched to a degree the community deems satisfying. Challenges have to start somewhere and should be given a chance to build their case gradually. These two cases, especially the AoE case, are (still?) a far cry from the comprehensiveness and precision of the alternative models and explanations we discussed and smack a little too much of *sui generis* challenges to established theories produced by coincidences. As such, they stand in stark contrast to Dirac's challenge to the orthodoxy and its explanation of the CMB, as he used coincidence as the core of a coherent alternative model, not a tool of criticism. Moreover, the alternatives we discussed dealt with an undisputed observational phenomenon in order to offer explanations, while in this case, the nature of the observational phenomena at stake is not firmly established. In fact, the observational context is much more similar to the erroneously detected deviations in the spectrum of the CMB in the 1970s and 1980s that we discussed, but the explanations picking up on deviations in the present case do not fit a more comprehensive view or model, thus decreasing their credibility. Having said that, the challenges like these two and moderate debates around them are certainly an indication of an appropriate playing field. Given the open-endedness of exploration in cosmology predicated on protracted underdetermination, especially the nature of the evidence that can be suggestive of alternative explanations even at a visual level, an absence of such challenges would be somewhat worrying.

29

The "problem of other observers" and anthropic reasoning

In several places throughout the discussion of alternative explanations of the CMB phenomenon, we have encountered anthropic reasoning deeply rooted in the epistemic structure of modern cosmology. It deserves a careful historical and philosophical analysis. While this is a huge topic in and of itself and, as such, cannot be accommodated here, in this brief chapter, we offer several entry points for such an analysis that hopefully will be forthcoming.

In this book, we use what could be called a "minimal" definition of anthropic reasoning, one corresponding to the Weak Anthropic Principle as categorized by Carter (1974). In other words, heeding the philosophical primer given by Nick Bostrom (2002), we construe anthropic reasoning as a set of observation-selection effects or biases related to our existence as evolved intelligent observers. This is usually applied to those physical and cosmological parameters that have unknown or poorly understood causes. For instance, at least prior to the advent of inflationary cosmology, we did not understand why the value of Ω is about 1, within an order of magnitude, but we could use anthropic reasoning to state that Ω cannot be, for instance, equal to 150, since such a universe would have re-collapsed very soon after the Big Bang, without sufficient time or other pre-requisites to evolve life and intelligence. That was exactly how Dicke (1961) argued against the large-number numerology of Dirac.[70]

Such usage immediately precludes all complicated nonsense ascribing a particular teleological role to anthropic reasoning. On the contrary, there are all kinds of reasons to regard anthropic reasoning as *dis*teleological, and these have been discussed in the literature.[71] By demonstrating how evolved complexity might *mimic* design, anthropic reasoning actually affirms disteleology and methodological naturalism, and this, we may speculate, is the source of much confusion in scientific, philosophical, and popular literature.

Some of the key anthropic papers were written by Hoyle et al. (1953), Dicke (1961), and Carr and Rees (1979), that is, persons intimately involved with most of the orthodox as well as the alternative explanations of the CMB. This did not prejudice their attitudes toward this approach at all: if anything, Hoyle sharply criticized anthropic reasoning (somewhat unfairly) in *Intelligent Universe* (Hoyle, 1983).

Most of the debates we have considered in this book occurred from the 1960s to the 1980s, significantly predating the resurgence of mainstream interest in anthropic reasoning that followed the formulation of eternal/chaotic inflationary scenarios (and *a fortiori* significantly predating the currently popular "anthropic landscape" of string cosmologies). Due to the timeline, therefore, anthropic arguments were as inconspicuous as possible, almost afterthoughts. The exceptions here, as elsewhere, were Hoyle's anthropic arguments. The fact that anthropic reasoning was masked, so to speak, should not deter us from trying to place it in an adequate methodological and historical context; in a sense, we need to "unmask" it.

Perhaps the most obvious and hence rarely discussed application of anthropic reasoning is justification for some kind of violation of the cosmological principle. For instance, in Ellis, Maartens, and Nel's (1978) model, we see the CMB as almost isotropic because we are far enough from the enclosing global singularity. Such a position, it is implied, is a necessary prerequisite for the evolution of life and intelligent observers like us. Here, the chain of reasoning is short and obvious: astrophysical locales near the singularity would be presumably exposed to much stronger physical stresses hostile to stability and other desiderata for abiogenesis and subsequent evolution. Therefore, to be able to discuss the issue in the first place, we have to assume a particular kind of evolutionary history, which is enabled only by narrowly selected properties of the physical environment. A similar situation can be found in Hoyle's 1975 proposal of zero-mass surfaces as the places of thermalization for the CMB photons: we are far from those surfaces, presumably because vanishing masses of elementary particles are not conducive to life and observability as we know them.

A somewhat more complex relationship holds for those alternatives that involve either primordial chaos (Rees, Eichler, Zel'dovich) or specific kinds of Population III objects (John Barrow, Hayakawa, Carr, etc.), or the Cold Big Bang (Layzer and Hively). These scenarios include significant modifications of the standard picture of structure formation and star formation history that may or may not suppress the chances of observers like us evolving in our epoch. It is very difficult to analyze the precise impact of each scenario, because the "galactic ecology" that regulates chemical and planetary evolution is highly nonlinear and likely chaotic.

Take for instance the Cold Big Bang "heresy." In an oft-cited but poorly understood study, Aguirre (2001) identified a Cold Big Bang model that is habitable according to the standards set by our terrestrial kind of life. This type of universe supports structures such as galaxies and stars, as well as chemical building blocks of life, with sufficient time for various kinds of evolution. Some opponents of anthropic reasoning hailed this result as a long-awaited refutation of the alleged "anthropocentrism" of such reasoning (e.g., Smolin, 2004). This conclusion just reflects deep confusion about what anthropic reasoning really means, however. Aguirre had, in fact, discovered another habitable region in the parameter space that overlaps with the region of the parameter space representing all possible Cold Big Bangs. This is an outstanding affirmation of anthropic reasoning, not a refutation, while it shows that we cannot just discard Cold Big Bang models on the basis of our existence – although nobody has actually claimed that, to the best of our knowledge.

As usual, we need to investigate explanatory hypotheses on a case-by-case basis. In contrast to the Cold Big Bang models, all – or almost all – steady-state theories, including closed steady-state ones like Ellis, Maartens, and Nel (1978), have a deep-seated conflict with anthropic reasoning, as first noticed by Paul Davies and subsequently elaborated by Frank Tipler.[72] As Davies (1978, 337) writes in relation to the Ellis et al. theory, "There is also the curious problem of why, if the universe is infinitely old and life is concentrated in our particular corner of the cosmos, it is not inhabited by technological communities of unlimited age" (see a discussion of this point in Barrow & Tipler, 1986). This is applicable to the classical steady-state theory as well, since Bondi and Gold's version and Hoyle's version actually contain galaxies of arbitrarily long age. Hence, the fact that we observe the "great silence" – or a generalized form of Fermi's paradox – is quite a bit more puzzling in such theories than in the relativistic Big Bang ones; in the latter, it is always possible to stipulate that there has not been enough time for some cosmologically visible anomaly due to intelligence and technology to occur.

Hoyle, for one, was keenly aware of this issue. If something is possible at any epoch, then by virtue of the "perfect cosmological principle," it has to be possible at any other epoch, including right now. The way out he chose in *The Intelligent Universe* represents a bit of a mystical cop-out: if we do not observe artifacts of arbitrarily old supercivilizations, this just means they are all around us, and we cannot distinguish them from "natural" entities and processes.[73]

Finally, we should mention that some of the problems with the *orthodox* CMB interpretation, like the cosmic variance problem, stem from what is essentially anthropic reasoning. We have just one map of the CMB anisotropies – because we are observers located in a particular location at a particular epoch. A natural conjecture, following from our Copernican prejudices, is that other observers observe a CMB map that although different in detail is statistically the same. Obviously, we cannot empirically test this, because it would require intergalactic travel over cosmological distances – the same dilemma encountered by early cosmologists like Einstein, Willem de Sitter, Edward Milne, or Sir Arthur Eddington when using "metaphysical" principles such as the cosmological principle.

Are our CMB observations truly typical then? Even if we don't address this extremely complex question at the intersection of fundamental physics and philosophy, we still need to acknowledge that this *is* a question properly falling within the scope of anthropic reasoning. People may not like the label, but while the radical social constructivist nonsense may argue otherwise, changing the label will change absolutely nothing.

30

The nature of boundary conditions in cosmology, the CMB, and the "laws of nature"

Several authors, starting with the philosopher Lucretius (1997) living in the first century A.D., have suggested the existence of the laws of nature, that is, the inevitable natural regularities, implies the finite nature of the universe. The laws had to have an inception date as the universe is essentially a winding-down mechanism; the fact it's still running means it was wound up at some point in the past (Kragh, 2017). This sort of argument was repeated in Isaac Newton's laws of motion. In a more sophisticated manner, Immanuel Kant (1797) argued in favor of an evolution of the universe governed by the physical laws formulated by Newton. He explicitly linked a cosmological view to the evolution of "new worlds." Another boost for the finite nature of the universe in light of the physical laws came with the discovery of the second law of thermodynamics (Kragh, 2017, 12). Had the universe existed forever, the heat death would have reached its maximum by now. (More recently, Roger Penrose [1989] suggested the time arrow in the universe is due to specific initial conditions in the creation of the universe and not to any specific physical laws that are time-reversal invariant.)

A more complicated but ultimately more justifiable notion of the impact of laws on the cosmological structure and dynamics came with the advent of atomic and quantum physics. Lemaître (1931), who first proposed the Big Bang scenario based on Einstein's equations, thought the initial singularity could not be grasped by the physical laws, as it implied the non-existence of space. This went hand in hand with a prevailing understanding of the breakdown of General Relativity in singularities (See e.g., Smeenk & Ellis, 2017). Yet quantum theory allows for fluctuations of the vacuum, and these were soon understood as the probabilistic laws that enable the origin of the universe. The quantum laws are probabilistic laws that can play the role of an initial cause of the evolving universe. The primordial quantum vacuum can be described by now ordinary quantum physics. Yet the idea that the relevant quantum laws provided by the quantum field theory pre-exist our universe (Ellis, 2007, 48) is only an assumption that leads to deeper questions we discuss below.

The notion of the laws also comes up in relation to the idea of the multiverse, which has been controversial, at least until very recently, because of its assumptions about causality. But it is also controversial in terms of the physical laws that govern it. Each world in the multiverse may harbor its own vacuum density, making the causal contact between these worlds even more problematic. Moreover, it is not clear whether this implies locality of the laws of nature, or whether there are unknown overarching laws.

This brings us to the question of the origin of the laws of nature that seems to be intricately tied to the origin of the universe. Unsurprisingly, an underlying question on how the laws of nature and cosmological knowledge, including existing cosmological models, should fit together was at the heart of dilemmas on the nature of the CMB among those outside the Big Bang and the general relativistic paradigm. Various alternatives to the orthodox view resulted in a trade-off between very different general assumptions of what should constitute cosmological knowledge.

In the spirit of the Copernican principle that states that our place of observation is not special, both stronger and weaker versions of the cosmological principle have been embraced in modern cosmology. On the one hand, it speaks to the breadth of the alternative explanations that such a fundamental principle was challenged by an explanation of the CMB based on George Ellis's elaborate cosmological model (Chapter 21). On the other hand, the authors of the alternatives based on the steady-state model and its modifications were actually very conservative in that respect. One of the motivating reasons for developing their alternatives was the pre-emption of the preposterous (in their view) possibility of a gradual change of the physical laws in time lurking in the background of Big Bang models.

Another related motivational reason for those in the steady-state camp to doggedly pursue alternative explanations of the CMB origin was tied to the inherent contingency of the initial conditions in the Big Bang scenarios. The crucial initial conditions determining all the major physical and cosmological parameters are in accord with regular physical laws, and the evolution of the universe can be studied all the way back with their help, but the initial conditions could have been different. A deeper worry about the method of the Big Bang models underlies this initial worry. The contingency of the initial conditions of a unique evolving universe implies the laws of physics encompassing all possible universes are not accessible even in principle, since it is impossible to experiment on a unique event, and experimenting (varying and controlling parameters) is ultimately the only reliable scientific way of discovering natural laws (Ellis, 2007). Thus, the regular attitude toward astrophysical laws (solutions of equations) as generally applicable to both observed and unobserved objects has to be abandoned at the cosmological level (McCrea, 1953). Even if there are underlying cosmological laws that determine initial conditions, it is not clear how we could test them even in principle as Ellis (2007, 49) suggests.

These worries are not merely abstract but are tied to the explanations of the CMB and reveal a deeper epistemic rift between them. Thus, both the orthodox explanation and the Cold Big Bang variants, in effect, use the CMB to test the initial conditions, not the actual physical laws of the evolution. The alternatives outside the general theory of relativity paradigm and the closed universe alternatives treat it as a test of the regularity of locally well-confirmed laws, while those relying only on regular physical laws to explain it do not go even that far. Such alternatives avoid the puzzles stemming from the apparent failure of physical laws to determine the actual initial conditions and the inability to study them with regular scientific methods.

If empirically viable, these alternatives would have had a point in avoiding the path taken by the orthodox interpretation, as the assumptions they aimed to challenge lead to an even more puzzling question. If we accept the above-described tension between initial conditions and physical laws, this implies another open question on whether the laws of nature capture the actual event of the physical creation of the universe. This opens up the possibility that "effective physical laws may depend on the boundary conditions of the universe" (Ellis, 2007, 31), or even that the laws vary in time and space, as argued by Dirac and in various Dirac-inspired cosmologies. It is also possible that the laws can approach this event only asymptotically. As Kragh (2017, 27) points out, it is not clear whether this concern is legitimately ontological at all, or if we can understand it only in operational terms.

31

The CMB and the multiverse

Limits of scientific realism?

One of the interesting vignettes from the "great controversy" between the relativistic and the steady-state theories was the debate between Bondi and Gerald Whitrow, in the course of which Whitrow sought to undermine the steady-state theory by pointing to some of its perceived philosophical deficiencies (Whitrow & Bondi, 1954; Whitrow, 1962).[74] For example, within the theory of Bondi, Gold, and Hoyle, there are finite regions of space that are never in causal contact with each other. Essentially, the classical steady-state universe is equivalent to what has become known as the *multiverse*: a set of disjointed cosmological domains ("universes"). Whitrow regarded this as unappealing, if not entirely unpalatable. Although his challenge was not widely accepted – and many considered that Bondi won that debate – the sense of unease about causally disconnected patches of spacetime remained salient.

That was just the beginning, however. As discussed in a refreshing paper by Ellis and Brundrit (1979), *even standard Big Bang cosmologies* are equivalent to the multiverse, if the spatial cross-sections are flat or open. For classical Friedmann models without dark energy, this would correspond to $\Omega = 1$ (Einstein-de Sitter) or $\Omega < 1$ models, which have an infinite number of galaxies and therefore an infinite number of habitable planets, living creatures, and intelligent observers. Properties of infinity, coupled with fairly uncontroversial results such as the second Borel–Cantelli lemma (e.g., Shiryaev, 2016), imply that each of us has an infinite number of exact or *almost exact* (in the technical mathematical sense) copies scattered throughout the infinite universe. Since cosmological horizons prevent us from causally interacting with those copies, the issue is metaphysical, rather than physical – but it is entirely possible that some will find it unpalatable in the same manner as Whitrow did.[75] At the very least, it seems issues like these exert some pressure – or perhaps impose a boundary – on our usual notions of scientific realism in cosmology.

Such pressure has recently been visible in debates about the multiverse emerging from chaotic or eternal inflation. Within the context of the inflationary paradigm, it turns out that the multiverse emerges in a generic manner. Quantum fluctuations in the inflation field (or fields) produce changes in the rate of expansion. Some regions expand slower, and others expand faster. Those regions with a higher rate of inflation expand faster and dominate the entire manifold, despite the natural tendency of inflation to end in other regions. This allows inflation to continue forever *in a set of regions of finite relative measure*; this leads to the conclusion that it is unnecessary to fine-tune the potential of the scalar field in order to achieve the end of inflation (the so-called "graceful exit" problem that plagued earlier versions of the theory).

This crucial insight is mostly due to Andrei Linde whose 1983 paper was revelatory: "This suggests that inflation is not a peculiar phenomenon that is desirable for a number of well-known reasons ... but that it is a natural and may be even inevitable consequence of the chaotic initial conditions in the very early universe" (Linde, 1983, 180). This is a great Copernican zooming out of our cosmological view. Since inflation happens somewhere at any given epoch, instead of "bubble universes" all nucleating and coalescing at about the same time as in Guth's "old inflation," many such bubbles *coexist* with the regions where inflation proceeds. The fact that inflation ended in *our* causally connected domain ("the universe") about 13.7 billion years ago tells us little about the wider whole; this wider whole is nothing other than the ensemble of all such causally connected cosmological domains, plus the de Sitter inflationary space. It is the inflationary multiverse.[76]

So it seems resistance to the multiverse is futile. In 2007, a collection of papers edited by Bernard Carr, *Universe or Multiverse?*, was published by Cambridge University Press. A symbolic milestone, it represented mainstream acceptance and even codification of this new view (Carr, 2007). The key point for a historian of science is that even those represented in the volume who are skeptical of the multiverse have accepted it as a legitimate field of scientific discourse, in sharp contrast to the view of the multiverse as a "metaphysical monstrosity" that predominated until a couple of decades ago. In spite of the question mark in the title, most contributors agree, even if hesitatingly and with a lot of qualifications, that the multiverse is the only rational choice given our best scientific theories of the moment. A delightful irony is that even those who are specifically skeptical of inflation or string theory have come up with their own multiverse versions, notably Lee Smolin and his self-reproducing universe (Smolin, 1997).

The new majority opinion, if not yet a consensus, was encapsulated by Sir Martin Rees as follows: "It is not absurd or meaningless to ask 'Do unobservable universes exist?,' even though no quick answer is likely to be forthcoming. The question plainly cannot be settled by direct observation, but relevant evidence can be sought, which could lead to an answer" (Rees, 2007, 61).

The history of multiple CMB explanations is rich with crossovers with the multiverse concept, beyond the personal connections of many of the actors, such as Carr or Rees. Hoyle's 1975 hypothesis on domains of opposing signs of mass is a novel and original form of the multiverse – where variable masses of particles are separated by null-mass boundaries, and the CMB is the main evidence of the existence of the boundary. Somewhat more expected are associations of, say, the cyclic universes, both re-collapsing and bouncing types, with the multiverse. After all, as many researchers have noticed (e.g., Barrow & Tipler, 1986) postulating very large or infinite series of universes successive in time is both epistemically and ontologically equivalent to postulating an ensemble of topologically disconnected universes in space.

There is an even deeper association of the CMB with the multiverse: today, we use it to test for predictions of chaotic inflation, in particular, the formation, collisions, and coalescence of "bubble universes." While the idea that early processes governed by quantum field theory at extremely high energies within the first minuscule fraction of a second not only create other universes but also leave potentially observable traces sounds too fantastic even by the standards of modern science, it is nevertheless taken quite seriously by theoreticians and even some observational astronomers. For examples from this domain, see Aguirre and Johnson (2011), Feeney et al. (2011), or Salem (2012).[77]

And there is more. The underlying idea of using the CMB as a kind of grand probe of cosmology taken in the loosest and widest sense has motivated even "crazier" studies venturing into metaphysical aspects of cosmology. Some have suggested using the CMB to investigate whether we are ourselves living in a numerical simulation (Hsu & Zee, 2006), presumably much more advanced, but along the lines of our own cosmological simulations discussed in Chapter 27. Note that both the theory *and* the phenomenology of the CMB are utilized in this approach. And the CMB is not a unique tool in this regard – a recent study suggested lattice QCD (quantum chromodynamics) physics for the same task (Beane, Davoudi, & Savage, 2014). Clearly, these approaches rely on the orthodox interpretation of the observed phenomena and open up a goldmine for philosophy of science.

In 2022, a massive compendium of difficulties with the current physical cosmology was published by Jim Peebles, Nobel laureate cosmologist and key architect of the orthodox CMB interpretation.[78] Many of the items mentioned are related in one way or another to the CMB phenomenology; perhaps the most significant are discrepancies between the CMB dipole anisotropy and the analogous dipole anisotropies in the distribution of other presumably cosmologically distributed sources, like rich galaxy clusters, quasars, and radio galaxies (Peebles, 2022, 18–24). Even perusing Peebles's encyclopedic review leaves the impression that the solidification of the orthodoxy was premature. He seems to have been (surprisingly?) open to anthropic reasoning and (surprisingly?)[79] open to the possibility that some new physics manifests itself on the cosmological scales.

In conclusion, while many scientists, together with historians and philosophers of science, will undoubtedly regard the multiverse theories as prototypical of "higher speculations" (cf. Kragh, 2011), our venture into historical experience suggests all radical breakthroughs encountered such reactions, and ideas ought to be weighed on their intrinsic merits, not on how they infringe on our cherished assumptions. This Enlightenment attitude underpins our continuous progress, and it is in the spirit of the Enlightenment that the complexity of real nature is acknowledged and appreciated. Nowhere in the domain of physical science is this better documented and clearer than in studies of the real and unembellished pathway that led us to understand the CMB and the physical universe.

Appendix A

Relativistic cosmological models

The classical textbooks such as Peebles (1993), Peacock (1999), Islam (2004), or Weinberg (2008) give the best overview of physical cosmology. Here, we summarize only a few concepts and formulas necessary for following closely the narrative of most chapters while perfectly aware that such treatment may look cartoonish at a glance.

According to the old saying, which is controversially ascribed to Eddington, Gamow or Allan Sandage cosmology is a "quest for three numbers," namely Hubble constant (H_0 or h), cosmological matter density (Ω_m) and the cosmological constant/dark energy (λ or the equivalent density Ω_λ). These three quantities determine which of the innumerable relativistic world models suggested in the literature may even become a candidate for approximate description of reality.

The most important dynamical variable in all cosmological models is the scale factor $R(t)$, which describes the change in global length scales with cosmic time, or how the distance of two random observers changes as a consequence of the Hubble expansion. The scaling factor is unobservable directly, but it is well represented by the cosmological redshift, which satisfies the following equation:

$$1 + z = \frac{R(t_0)}{R(t_1)},$$

where t_0 is the epoch of observation (present day!) and t_1 is the moment of emission of the signal (presumably light or any other electromagnetic wave). Putting $t_1 = t_0$ immediately shows why the entire local universe – for example, sources within the Local Group or in the nearby groups of galaxies – are often labeled in cosmological texts as $z = 0$ region.

The scale factor plays the main role in the most general form of spacetime metric compatible with the cosmological principle, which is the Robertson–Walker, or sometimes Friedmann–Robertson–Walker (FRW) metric. In spherical coordinates (r, Θ, φ) it has the canonical form:

$$ds^2 = c^2 dt^2 - R^2(t)\left[\frac{dr^2}{1 - kr^2} + r^2\left(d\theta^2 + \sin^2\theta \, d\phi^2 \right) \right].$$

Here, k is the discrete parameter that discriminates between open ($k = -1$, hyperbolic), flat ($k = 0$, Euclidean) and closed ($k = +1$, elliptical) global geometries. The scale factor $R(t)$ is usually obtained by solving the Friedmann equations for various kinds of matter and their corresponding equations of state.

The scale factor could be expanded in Taylor series about any given moment of cosmic time, thus including the present moment t_0. Starting with small temporal differences $t_0 - t$, we obtain:

$$R(t) = R\left[t_0 - (t_0 - t)\right] =$$
$$= R(t_0) - (t_0 - t)\dot{R}(t_0) + \frac{1}{2}(t_0 - t)^2 \ddot{R}(t_0) - ... =$$
$$= R(t_0)\left[1 - (t_0 - t)H_0 - \frac{1}{2}(t_0 - t)^2 q_0 H_0^2 - ...\right].$$

Here, the definition of the Hubble "constant" is given as:

$$H_0 = \frac{\dot{R}(t)}{R(t)}_{t=t_0} \quad ;$$

it has units of $\left(\text{time}\right)^{-1}$, with the reciprocal representing the Hubble time, within the factor of the present-day age of the expanding universe. The dot denotes differentiation with respect to the cosmic time, and q_0 is the (present-day) *deceleration parameter*.

In practice, the Hubble constant is often parameterized as

$$H_0 \equiv 100\, h \text{ km s}^{-1} \text{ Mpc}^{-1},$$

where h is the dimensionless quantity roughly between 0.5 and 1. *WMAP* results gave the best value for this parameter as (Spergel et al., 2007):

$$h = 0.732^{+0.031}_{-0.032}.$$

While the deceleration parameter is in the general case defined as:

$$q(t) \equiv -\frac{R(t)\ddot{R}(t)}{R^2(t)}.$$

In contrast to the scale factor, the deceleration parameter q_0 is an observable, dimensionless quantity. The goal of most of the neoclassical tests of world models (Peebles, 1993) was empirical determination of this parameter. In terms of dimensionless cosmological densities,[80] a simple relation holds:

$$q_0 = \frac{1}{2}\Omega_m - \frac{c^2\Lambda}{3H_0^2} =$$
$$= \frac{1}{2}\Omega_m - \Omega_\Lambda.$$

Negative values of the *deceleration* parameter mean that the universe is actually accelerating, which has been discovered by observations of the distant Type Ia supernovae in 1998. This was highly surprising, not least because most of observational cosmology studies in 1980s and 1990s suggested positive q_0, mostly between 0 and 0.5, but also sometimes going to absurd values as high as 7.9 from optically variable QSOs (Cheng, 1981). Cosmological density fractions for matter and dark energy are:

$$\Omega_m \equiv \frac{\rho_m}{\rho_{crit}} = \frac{8\pi G}{3H_0^2}\rho_m;$$

$$\Omega_\lambda \equiv \frac{c^2}{3H_0^2}\lambda.$$

Here ρ_m is the average physical density of matter (in g cm^{-3}) in the universe at a given epoch, usually present-day. Notice that for static universe Hubble constant would be zero (and Ω_m would formally tend to infinity), therefore it makes no sense to talk about cosmological densities in such cases; which is the reason why Einstein's 1917 static universe is not listed in Table A.1.

Finally, a few words on the concept of cosmological horizon are in order (again, many important presentations are to be found in the literature). Cosmological horizon is, just like the terrestrial horizon, the distance up to which we can see – roughly speaking, the radius of our visible universe. More precisely, in universes for which the cosmological principle is valid, this is the characteristic length scale over which causal influences may propagate. Although there are multiple kinds of cosmological horizons, the most basic type is the particle horizon given, at the current epoch t_0, as

$$D_h = R(t_0)\int_0^{t_0}\frac{cdx}{R(x)}.$$

Even without knowing the exact functional form of the scaling factor $R(t)$, from this formula we clearly see why horizon size will be very small on the CMB map – hence why the horizon problem of pre-inflationary cosmology is so serious. When t_0 decreases, *two* things contributing to decreasing size of the horizon happen simultaneously: both the upper limit of the integration decreases (shorter range of a causal signal) and $R(t_0)$ goes down (space contracts). The synergistic effect of the two explains why, at the surface of last scattering, the causally connected patch subtends only about 4 square degrees in the sky. It is only about 1/9,000 of the full CMB map (which, of course, subtends 4π steradians), demonstrating how remarkable the uniformity of the CMB signal up to the $\Delta T / T \sim 10^{-5}$ level of the small-scale anisotropies really is: one would be hard-pressed to find anything else in nature that behaves *uniformly over the equivalent of 9000 visible universes!*

Table A.1   *Some relativistic cosmological models with corresponding densities and deceleration parameters.*

Model	Ω_m	Ω_λ	q_0	Comment
de Sitter	0	1	-1	"Empty" universe expanding exponentially with $R(t) \propto \exp(t)$, but also leading to the same expansion dynamics as the classical steady-state model.
Einstein–de Sitter	1	0	0.5	The simplest, most symmetric and stable case of matter-dominated universe, still often used for pedagogical purposes.
Observed open universe	0.05–0.3	0	0.025–0.15	Almost all observations ca. 1950–1998 suggested this group of models, still requiring most matter to be dark.
λCDM	0.3	0.7	-0.6	Asymptotically approaches the de Sitter exponential case.

Note that all cases in the table represent geometrically flat universe, except for the "observed open universe" (which has a hyperbolic geometry).

Appendix B

Dipole anisotropy

The dipole anisotropy – whereby one pole of the CMB is slightly cooler than the other – is one of those physical phenomena that are conceptually simple (at least at the first and even second glance), yet immensely rich in terms not only of knowledge and explanation, but also on the methodological level. All those popular examples like running or driving in the rain testify on how widely present is the desire to have an accessible, laypeople-friendly explanation of a piece of supposedly arcane and exotic cosmological discourse.

The dipole anisotropy was discovered by Edward Conklin of Stanford University in 1969, and was confirmed by many subsequent surveys, both ground-based, air-based (balloons and planes), and ultimately satellite observations. Our understanding of dipole anisotropy on both theoretical and observational level solidified in the late 1970s and early 1980s. The work of Brian Corey then at MIT and the Berkeley group headed by the future Nobel laureate George Smoot (and in part also the great nuclear physicist, Luis Alvarez) was decisive in this regard. The Lockheed $U-2$, then the highest-flying airplane in existence, was used as one of the first flying observatories due to Smoot's political and organizational acumen. For some of the historical aspects see Peebles et al. (2009); also Kragh and Longair (2019).

The explanandum is usually expressed as the measured fact that the observed CMB dipole is about a hundred times larger than the higher multipoles. The latter are ascribed to inherent physics of the surface of last scattering (such as the Sachs–Wolfe effect) in the standard CMB interpretation, with a modest admixture of later-epoch events happening to photons while traveling from the surface of last scattering (such as the Sciama–Rees or Sunyaev–Zel'dovich effects). The dipole, however, cannot be explained by any of those cosmological processes, as long as the standard symmetries are accepted.

The *simplest* interpretation of the observation that the dipole is so much stronger than the higher multipoles is to assume that the solar system barycenter moves with a velocity of 369.82 ± 0.11 km/s toward the Galactic coordinates $(l, b) = (264.021° \pm 0.011, 48.253° \pm 0.005)$ with respect to the surface of last scattering. The surface of last scattering is assumed to define the frame of reference in which CMB temperature is isotropic (up to the small-scale inherent anisotropies). This motion causes a standard Doppler shift, blueshift in the direction of motion and redshift in the opposite direction. This has been occasionally dubbed the "kinematic dipole" in the literature. This velocity has not only been measured with high accuracy, but the Doppler effect has also tentatively been confirmed by its second-order effects (aberration and modulation) on higher multipoles (da Silveira Ferreira & Quartin, 2021). It is also the only explanation that does not invoke any exotic physics or allegedly observable consequences of unobserved and necessarily speculative phenomena.

The magnitude of the dipole strength can be obtained from the general decomposition of the CMB map in terms of spherical harmonics in the direction given by the unit direction vector $\vec{x} = (\vartheta, \varphi)$:

$$T(\vec{x}) = \sum_{\ell=1}^{\ell_{max}} \sum_{m=-\ell}^{\ell} a_{\ell m} Y_{\ell m}(\vec{x}).$$

Taking into account only the monopole and the dipole terms, we obtain the classical, pre-*COBE* picture of almost-uniform celestial background with one large anisotropy. Dipole corresponds to the $\ell = 1$ term, for which the (orthonormalized) spherical harmonic reduces to the cosine function:

$$Y_{10}(\vartheta, \varphi) = \frac{1}{2}\sqrt{\frac{3}{\pi}} \cos \vartheta.$$

The other two spherical harmonics in the dipole sum, Y_{1-1} and Y_{11} are of equal absolute value and opposite sign, so they cancel out. Here and elsewhere, θ is the angle between the direction of motion and the direction of observation. Overall, the ratio of dipole to monopole is given as:

$$\frac{\delta T_{\ell=1}}{T} = \frac{v}{c} \cos\theta, \tag{*}$$

in the simplified *classical* picture. When observed in the direction of motion, (*) reduces to the famous simplified formula:

$$v = c\left(\frac{\delta T_{\ell=1}}{2.7 \text{K}}\right).$$

It has been used in this form by, for instance, Cheng et al. (1979), who in their Table 2 (p. L142) list other important measurements on the Doppler velocity inferred from the dipole, including the early survey of Partridge and Wilkinson (1967), which could only place the upper limit on the dipole strength. In retrospect, it is clear that Earth-based (including balloon- and plane-based) surveys have been imprecise enough, especially in declination, ranging from $-30°$ to $+19°$ (cf. Fabbri et al., 1980); the modern *WMAP/Planck* value is about $-6.9°$.

The magnitude of the dipole is (modern *Planck* value):

$$\delta T_{\ell=1} = (3.3621 \pm 0.0010) \times 10^{-3}\,\text{K}.$$

It is usually quoted as about 3 millikelvin, which is conveniently about thousand times smaller than the monopole.

In the full relativistic treatment of the Doppler shift (e.g., Weinberg, 2008, 129–130), we obtain:

$$T + \delta T_{\ell=1} = \frac{T}{\gamma\left(1 + \dfrac{v}{c}\cos\vartheta\right)}.$$

Where the relativistic gamma factor is, as usual

$$\gamma \equiv \frac{1}{\sqrt{1 - \dfrac{v^2}{c^2}}}.$$

(Everywhere here, v is the speed along the direction of the motion, that is, the maximal Doppler velocity. Note that for photons coming directly from the direction of motion one has $\cos\vartheta = -1$.) Solving for velocity obtains:

$$v = \frac{c}{\cos\vartheta}\frac{1}{1 + \dfrac{\delta T_{\ell=1}}{T}}\left(\frac{1}{\gamma} - 1 - \frac{\delta T_{\ell=1}}{T}\right).$$

Among alternative explanations, one should mention the "tilted universe" toy model of Michael Turner (1992) invoking strong isocurvature perturbation at large (superhorizon) scales. In this model, there is a large-scale streaming motion of matter caused by this perturbation – this would be a special kind of peculiar velocity, or indeed *peculiar* velocity. This model has a curious property that the universal rest frame and the frame in which CMB looks isotropic (apart from small-scale intrinsic anisotropies) do not coincide: the material universe seems *tilted* toward the center of mass of the isocurvature perturbation, similar to the way of all kinematic phenomena may appear tilted in a house built parallel to the ground on a gentle mountain slope. Other attempts have been made about the same time. Notably, Paczynski and Piran (1990) have considered off-center observers in a spherically symmetric Tolman–Bondi universe where offset from the center could simulate the dipole anisotropy. Obviously, all such solutions violate the cosmological principle and hence are mostly important for constraining the impact of local density fluctuations.[81]

The main underlying physical lesson of the dipole anisotropy is that its measurements (especially the *COBE FIRAS* ones) have decisively confirmed the reality of large peculiar motions of galaxies and galaxy groups. Peculiar velocities of galaxies in the vicinity of the Milky Way have been a subject of long-standing controversy since the times of Hubble and Sandage. Although we cannot discuss this fascinating subject here, it is important to notice that early intuitions of Vera Rubin have been confirmed both by studies of nearby elliptical galaxies (the famous "Seven Samurai" collaboration; Dressler et al., 1987 and subsequent papers in that series) and by precise measurements of the dipole anisotropy. Since the local map of peculiar velocities pointed toward the existence of the "Great Attractor" and the Local (Laniakea) Supercluster – it has been exactly the consistency and concordance of these results that created not only high confidence in the standard cosmology, but also an impressive confirmation of the correctness of a wider astrophysical picture of the architecture of our universe.

In turn, this brings us into contact with the issues of global cosmological symmetries, the cosmological principle of Eddington and Milne, as well as the most general issue of underdetermination of our cosmological theories. This has been subject of much recent philosophical work (e.g., Beisbart, 2009; Butterfield, 2014), which we cannot seriously discuss here. What needs to be mentioned, though, is that both the principle itself and its relationship to other topics of astrophysical research, notably the studies of the large-scale structure, have remained an active research area in both theoretical and observational sense (e.g., Nadolny et al., 2021).

Accounting for the dipole anisotropy has been a task for *any* explanatory hypothesis for the CMB origin. This instance highlights another feature of contingent, "bush-like" nature of scientific progress: a particular feature may sometimes be regarded as more or less isolated from the overall theoretical and explanatory framework. Clearly, we should not expect alternative explanations of the CMB origin to follow the standard explanation verbatim; however, each and every hypothesis that assumes the validity of large-scale homogeneity and isotropy is likely to adopt the very same interpretation of the dipole anisotropy. This clearly applies to most hypotheses we have discussed in Part V; only radically inhomogeneous models, like those of Ellis et al. (1978) or Phillips (1994a, 1994b), considered in Chapter 21 are required to provide a specific explanation for this particular feature of CMB. The "tilted universe" has some similarities with the Phillips model, both invoking large-scale streaming of matter. Otherwise, there is no reason to suspect that, for example, hypotheses postulating dust-grain thermalization as the origin of the CMB photons reject peculiar motions of galaxies or, even more radically, reject the reality of classical Doppler effect, therefore requiring a special explanation for the observed dipole.

In other words, the dipole anisotropy is a part of the standard comparison set or a "control group" (in terms of conventional empirical research in life or social sciences) phenomenon: both the orthodox CMB interpretation *and* the alternatives are required to pass this base test of accounting for the dipole strength. The key insight is that the standard account is founded upon both conceptually and historically simpler physics and astronomy, notably the Doppler shift and the existence – for a long time controversial – of significant peculiar motions/large-scale structure. While the latter is related to cosmology, the relationship is indirect and involves many additional steps; it is much more complex than the relatively straightforward inference from the blackbody shape to the primordial fireball originating with Gamow, Dicke, et al.

Notes

Epigraph (p. V)

1. Heraclitus B8 DK, cited by Aristotle (EN 9. 2 1155b4–6). The fragment translated by Sandra Šćepanović.
2. Hoyle (1955).

Introduction (pp. 2–3)

3. The term "epistemological" typically pertains to the comprehensive examination and reflection on the nature of knowledge, encompassing scientific knowledge, that is, to epistemology. On the other hand, "epistemic" usually refers to one's concrete stance concerning a specific element of knowledge, such as an observation, theory, or a model, within the realm of science and beyond. While these terms may occasionally be used interchangeably, we generally maintain the distinction to differentiate between a broader philosophical inquiry into the nature of scientific knowledge, and more specific concerns and attitudes pertaining to particular contexts in scientific inquiry. Certain cosmologists, such as Hermann Bondi, explicitly formulated their own comprehensive perspective on scientific knowledge (of broader epistemological relevance) that also informed their scientific work. Some occasionally somewhat explicated their specific (epistemic) attitudes to their research, in order to justify their cosmological approaches. Others simply assumed particular pre-existing (epistemic) attitude to specific scientific research, without extensive reflection.
4. A more mundane yet telling example is the case of physicists at CERN who failed to detect a new kind of mesons that their apparatus was in fact producing in their experiments at the time due to their neglect of the recent theoretical account of that particle (Pais, 1986, 97; Perović, 2011).

1 Physical cosmology from Einstein to 1965 (pp. 7–8)

5. Einstein (1917). For the debate of Boltzmann and Zermelo, see Boltzmann (1895) and a beautiful introduction by Steckline (1983).
6. Olbers's paradox, often called the "dark sky paradox," refers to the fact that the universe is obviously far from the thermodynamical equilibrium: A small number of hot sources visible in the night sky are in disequilibrium with the cold dark space between them. This is incompatible with the naïve picture of an eternal infinite universe (see the classical discussion in Bondi, 1960, as well as a modern one in Peebles, 1993). Similarly, gravitational paradox suggests that

an eternal, static universe is incompatible with the Newtonian physics, since it would tend to collapse under the unbalanced influence of the gravity of its constituents (stars or galaxies). In both cases, postulating the finite age of the universe seems to open at least some wiggle room for resolution of the difficulty.

7. The label itself was highly unpopular and remained so (with a few notable exceptions) well into the post-1965 period; for an amusing and lively account of this, see Kragh (2015).

8. This has also been variously attributed to Eddington, Lemaître, or the German-British radio astronomy pioneer Peter Scheuer (cf. Longair, 1993).

2 The "great controversy" (1948–1965)
and epistemological issues it raised (pp. 10–12)

9. See, for example, Sulliven (1931).

10. See, for example, Gee (1999).

11. The locution is due to Peebles (1993).

12. There have been subsequent "small age problems" from time to time, which fueled dissatisfaction with the standard Big Bang orthodoxy in people such as Hoyle, Burbidge, and their allies: occasionally the best-fit age for an extreme subdwarf (like SMSS J0313–6708 whose metal content is less than 1 in 10 million of the solar one!) or a particularly old globular cluster (like Messier 15) would clash with the age of the universe in the currently popular Big Bang model. Especially egregious problems were created by the Einstein-de Sitter (EdS) model with $\Omega = \Omega_m = 1$, which has a very short age of $2/3\,H_0^{-1}$, where H_0 is the present-day Hubble parameter. All those minor age problems were alleviated, if not entirely and finally resolved, by the discovery of dark energy in 1998. Irrespective of its other properties, all conceivable forms of dark energy lengthen the cosmological age scale, thus enabling more comfortable margins for the formation of the oldest generation of stellar objects.

13. In observational astronomy, standard candles are sources with constant (and preferably *known* – although that stronger condition is not necessary for all applications) luminosity over space and time. The most famous standard candles in history of astronomy have been classical Cepheids – variable stars with regular pulsations, named after the prototype δ Cephei. Determinations of distances, especially large distances characterizing extragalactic astronomy and observational cosmology, heavily rely on finding appropriate standard candles.

14. λCDM means: λ (cosmological constant) plus cold dark matter (as the main component of matter in the universe and the main factor in the structure formation processes).

5 CMB phenomenology (p. 24)

15. It also gives rise to the 3-4-5 mnemotechnic device sometimes used in classrooms: relative anisotropies are characterized by consecutive orders of magnitude, namely the dipole anisotropy (10^{-3}), Sunyaev-Zeldovich and related effects (10^{-4}) and intrinsic anisotropies (10^{-5}).

6 Standard "textbook" history and its
shortcomings (pp. 31–32)

16. See also Weinberg (2008). Stebbins (1997) provides a detailed theoretical account of the CMB.

17. As in the case of many other unpopular (or, from another perspective, unfashionable) research topics, individual researchers or small research groups have either piggybacked work on the subject in some wider-area of institutional funding, or relied entirely on external grants less constrained by the orthodox thinking (e.g., the Templeton Foundation or FQxI grants).

7 Emergence of precision cosmology (p. 33)

18. This was also true of institutions where those who made a career in cosmology were educated. See, for instance, Fred Hoyle's account of his college circle in Gregory (2005, 13). Many distinguished cosmologists started their research careers in different fields (Gamow, Dicke, Peebles, Ostriker, Weinberg, Zel'dovich, etc.) even after it became formally and financially possible to declare oneself a cosmologist.

8 Underdetermination of theories and models in cosmology (pp. 42–47)

19. It remains to be seen whether this applies to the study of the origin of life.
20. Some authors argue that even fundamental physical theories, such as quantum mechanics, quantum field theory, or string theory, suffer from underdetermination in the form of competing interpretations (Belousek, 2005; Fraser 2009; Matsubara, 2013; Acuña, 2021), but it is not clear that this sort of uncertainty is of the same sort as that in historical sciences, since all or most empirical evidence has been available for some time in the former. In any case, we are interested in underdetermination due to the indirect nature of evidence.
21. Stanford (2006) offered his own examples of viable alternatives to key historical biological theories, but his goal was to provide examples of previously unconceived alternatives as an argument against naïve realism in science, quite apart for the goals of our study.
22. John Earman's 1993 paper preceded it. It was a very pointed case analysis rather than a general discussion of the type Butterfield pursued. Earman (1993) offered ample evidence of actual empirical equivalence of cosmological models allowed by Einstein's equations even in principle.
23. Insulin has 51 amino-acids. The error was corrected in a later related work by Butterfield (2014).

10 Classifying and analyzing unorthodoxies (pp. 57–59)

24. For present purposes and without downplaying the epochal importance of that project, we ignore attempts to go beyond general relativity in terms of *quantization* of gravity.
25. See Maher (1993) for the distinction of prediction and accommodation of experimental results.

11 Cold and Tepid Big Bangs: Population III objects (pp. 63–70)

26. Sometimes these models are incorrectly labeled Big Bang models with low entropy (e.g., Penrose, 1979), as there are deeper entropy reservoirs than the photons+baryons system (Egan & Lineweaver, 2010).
27. See Kragh (2013b) for the early history of these models.
28. As detailed by Gould and Vrba (1982), exaptation is an adaptation that has, over time, been coopted for other functions. Stock examples include wings, originally formed as adaptations for cooling the animal's body and subsequently *exapted* for a new function: flight.
29. For example, Komatsu et al. (2011).

13 Thermalization by grains, the first wave (pp. 82–84)

30. An interesting side question that did not confront the orthodoxy but emerged from the account concerned the cosmic rays as the early source of energy through gravitational collapse (Layzer, 1968, 99). The origin of cosmic rays as a cosmologically relevant issue occurred here and there in steady-state alternatives as well.

31. We should remember here and elsewhere that the conversion from redshifts to ages (or distances) occurs differently in different cosmological models. See Appendix B.
32. It is not clear why Layzer and Hively did not discuss this case (arguably less "bizarre" than the spherical shells) in some detail.

14 Primordial chaos (p. 87)

33. In particular, modern approaches based on Boltzmann's kinetic equation show that when the free electron fraction in primordial plasma falls below ~ 10^{-2}, photons decouple. Since the epoch of recombination is defined as the period of rapid decline of this fraction from 1 to about 10^{-3}, decoupling occurs during the recombination. It is still a useful exercise for advanced undergraduate or graduate students to derive the redshift of decoupling in the counterfactual case that recombination never occurred (Dodelson, 2003). As it turns out, it would have happened much later, at $z \simeq 40$, simply because the Hubble expansion makes it progressively more difficult for photons to find electrons to scatter off them.

15 Early intergalactic medium, massive Population III objects, and the large-numbers hypothesis (pp. 93–97)

34. The fact that Dirac's Nobel-prize-winning prediction of the existence of antimatter, based on the so-called "hole theory," including states with negative energy – as well as a more general fact that Dirac was the main pioneer of the quantum field theory – should perhaps count as indicative for the latter option, which does not violate the conservation of energy and shares some common ground with other established quantum field phenomena with negative energy states (Casimir effect, Hawking radiation of black holes, etc.).
35. We note here in passing that due to the slow expansion rate, Dirac's and similar models have a particularly drastic form of the age problem; it turned out impossible to raise the age of the universe to anything higher than about 6 billion years. In effect, any age measurement of any astrophysical system giving a higher age – there are plenty of them, in particular where the Milky Way globular clusters or halo metal-poor stars are concerned (e.g., Cowan et al., 2002) – can be regarded as clear falsification of these models.
36. For example, the paper of Canuto et al. (1977) has 237 citations in the NASA ADS database, and even more impressive score of 523 citations on Google Scholar (as of June 2023); by standards of most papers proposing CMB alternatives studied in this book, this is a performance significantly above average.
37. Tolman (1934); see also Weymann (1966) for the application to CMB.
38. Obviously, even more fundamental obstacle in front of Dirac-inspired theories is the experimental refutation of variability of *G*. While this is not related to CMB and hence beyond the scope of the present book, we just mention in passing that both experiments in the solar system with precision ranging of our interplanetary probes (Hellings et al., 1983) and observations of the binary pulsar as a natural "gravity lab" (Damour, Gibbons, & Taylor, 1988) have decisively shown that even if *G* is variable, it is so much less than required by Dirac's LNH. Hence, even in 1980s, before *COBE* and its high-precision constrains on the spectral distortion and chemical potential of CMB, the main rationale for this class of theories has evaporated.
39. NASA Astrophysics Data System (www.adsabs.harvard.edu/) lists 121 citations (as of June 16, 2023), corresponding to a moderately successful 2.75 citations per year. In comparison, the database lists 1.18 citations per year for Layzer and Hively's (1973) paper.
40. We should also note that in contrast to most unorthodox cosmological models discussed in this book, Dirac's cosmology has never entirely disappeared from the scene – a simple ADS, Web of Science, or Google Scholar search will demonstrate that there is a minuscule, but rather steady trickle of research on Dirac's or Dirac-like cosmologies and their crossovers with more modern topics like varying dark energy, exotic types of dark matter, or coupling of various hypothetical

scalar fields to gravity (e.g., Mercier, 2019; Moradpour et al., 2021). One may speculate about the perennial impact of the wider idea of variation of the fundamental "constants" and changeable laws of nature on the imagination of theoretical physicists and cosmologists (and its even wider cultural impact). This worthy topic, however, is beyond the scope of the present book.

16 Late thermalization of starlight (p. 99)

41. Again, it can never be overemphasized how important it is to properly place such assessments within the observational context of the times. We knew very little about the autocorrelation function of galaxies and the strength of its clustering prior to the key study of Davis and Peebles (1983); even less was known about higher-level structures like superclusters and voids. Therefore, astronomers of Rana's time tended to underestimate and downplay these deviations from the cosmological principle.

17 "An excess in moderation": High-baryon universe (p. 101)

42. NASA ADS lists 118 citations of the paper, corresponding to about 3.9 citations per year, at the high end of the alternative explanatory hypotheses considered in this book (last checked June 24, 2023). This might have been slightly helped by its position as the *first* paper printed in the November 20, 1992, issue of *The Astrophysical Journal*, especially in the 1990s when people still took pride in subscribing to the hefty white hardcopy editions of the premier research outlet in astrophysics and cosmology!

18 Motivations: Who's afraid of the Big (Bad) Bang? (pp. 105–107)

43. For details concerning geodesics, see Hoyle (1948, 376).
44. Ellis (1984) gives a detailed theoretical classification of radical cosmological unorthodoxies up to the early 1980s.
45. For a somewhat different assessment see Sorrell (2008).
46. One of the most cogent formulations is given by LaViolette (1986). For classical refutations, see Alpher (1962), Zel'dovich (1964), and Wright (1987); empirical arguments up to the mid-1980s were summarized in a comprehensive review by Sandage (1988); for a more recent negative test, see Foley et al. (2005).

19 Hoyle–Narlikar theory and the changing masses origin of the CMB (pp. 108–112)

47. This does not mean they endorsed orthodox cosmology either. After 1965 (if not before), both turned to other research topics outside cosmology and achieved remarkable success in fields as heterogeneous as gravitational waves and teaching of science (Bondi) or pulsars, geophysics, and biology of extremophiles (Gold). As Kragh aptly put it: "Without saying so directly, Bondi and Gold admitted that the battle was lost; but they did it with regret and without embracing the victor" (1996, 380).
48. One of the appealing features of any such scheme is that it automatically answers a philosophically interesting question that bothered Einstein, namely why, in contrast to electromagnetism, all charges to which gravitation is coupled (= masses) have the same sign.

49. This is not exactly true, as the surface of last scattering will be provided by whatever surface close to the zero-mass surface is capable of producing a spectrum deviating from the perfect blackbody because of instrumental uncertainties. This is not just splitting hairs. As a thermalizing agent, the zero-mass surface and its vicinity are stronger than any other such agent conceived by any other alternative hypothesis discussed here (with the possible exception of Davies's time-reversal hypothesis discussed in Chapter 21). Therefore, Hoyle's hypothesis cannot be rejected on the basis of insufficient thermalization, in contrast to Rowan-Robinson's or Rana's models.

50. Among 39 citations reported in the NASA Astrophysics Data System by June 2023 – a remarkably small number for a paper published 48 years ago, by such a famous author and on such a hot topic – one is a self-citation, one is a book chapter devoted to Hoyle's achievements, six (15.3%) are by a single author (Canadian cosmologist Paul S. Wesson, with collaborators) interested in the variability of fundamental physical constants, and three deal with the issue of the CMB origin; none of the latter was published after 1980. A mild rekindling of interest is related to the concept of the power law inflation (Paul, Sengupta, & Ray, 2023).

20 Revised steady state (pp. 114–117)

51. This could be derived from C-field dynamics; see, for example, Hoyle, Burbidge, and Narlikar (1993).
52. See also Kragh (2012).

21 Closed steady-state models (pp. 118–123)

53. This is refreshingly reminiscent of Alfred Russel Wallace who, in a book published in 1903 (!) argued that only near the center of his model universe – similar to what was then known as the Kapteyn universe – can life and human mind evolve (Wallace, 1903; see Ćirković, 2012).
54. See Bondi (1955, 1992) and comments in Kragh (1996).
55. Observational astronomers are accustomed to using *brightness temperature* as a measure of the intensity of radiation coming from a source. Brightness temperature is equal to the antenna temperature if the source is a sufficiently good approximation of a blackbody.
56. While Hoyle does not cite Davies's paper, he was surely aware of its existence. The lack of mention more likely reflects Hoyle's infamous stinginess with citations (his 1975 paper has only two references).

22 CMB in plasma cosmology (p. 125)

57. See www.astro.ucla.edu/~wright/lerner_errors.html, last accessed June 18, 2023.

24 History and epistemology: The emergence of orthodoxy (p. 133)

58. As mentioned in Section 3.1, Rowan-Robinson (1974) was an exception; he used Milne's special relativity cosmology rather than Friedmann models as the background. However, there was no particular motivation for this. He explicitly noted that the Milne model "is a good approximation to a low density $\Omega \approx 0$ Universe with matter" (Rowan-Robinson, 1974, 46), that is, an open Friedmann model. Rowan-Robinson, of course, never doubted the general Big Bang picture, either in the work referenced here or in other papers and books.
59. Again, with the exception of Ned Wright, whose highly informative web site (www.astro .ucla.edu/~wright/errors.html, last accessed March 30, 2024) contains important criticisms of the non-mainstream approaches.

25 What about the alternatives?
(pp. 138–142)

60. This is sometimes disputed, but a full analysis is beyond the present scope; see Disney (2000); Lopéz-Corredoira (2014).
61. See note 14 above.
62. There are many dangerous preconceptions in popular controversies of science and pseudo-science (e.g., those related to global climate change or universal vaccination), which could and should be easily dispersed by investigations of "science at work." Interestingly, the defenders of science often use misplaced and misguided arguments that are easily demol-ished by detailed analysis of the history or philosophy of specific case studies, such as we do here. For example, at least one influential web encyclopedia explicitly devoted to "[a] nalyzing and refuting pseudoscience and the anti-science movement" regards "alternative cosmology" as belonging to the same pseudoscientific category as "alternative medicine" (http://rationalwiki.org/wiki/Alternative_cosmology, last accessed July 31, 2023). Apart from the similar sound, there is no parallel between the two in either an epistemological or an ethical sense.
63. In this manner, the whole case study may be understood as a refinement of falsificationist views of scientific knowledge (Popper, 1972, 1992). The view has been taken seriously by astronomers and cosmologists (Kragh, 2013a).
64. "If it is not true, it is well conceived."
65. A tangential issue has been relevant to the attempts, mainly by Canadian astrophysicist Paul Wesson and his collaborators, to build a cosmological model of our universe embedded in a classical spacetime of higher dimensionality (e.g., Wesson & Seahra, 2001).

26 Pragmatic aspects of model building
and social epistemology of cosmology (p. 145)

66. See Perović (2018) for a review of relevant studies.

27 Large-scale numerical simulations in cosmology:
Beyond the theory–observations distinction? (p. 149)

67. Willman et al. (2005).
68. Einasto et al. (2016).
69. The distinction is related to the possibility of changing one's point of view. Barring some dra-matic science-fictional development like traversable wormholes, this will almost by definition need billions of years, thus putting a significant strain on the patience of anyone wishing to observationally test the matters involved.

29 The "problem of other observers"
and anthropic reasoning (pp. 157–159)

70. One fact is often ignored or swept under the rug, especially by vehement opponents of the anthropic reasoning with an axe or two to grind, such as Earman (1987), Pagels (1998), and Klee (2002): one could apply anthropic reasoning in fields far removed from fundamental physics and cosmology if the selection effect in question impacted the number and distribution of observers. Examples of this kind are widely dispersed over planetary science, ecology, risk analysis, and even economics and policy studies (Ćirković, 2016).

71. Notably Bostrom (2002); Barnes (2012). For teleological misuse of anthropic reasoning specifically pertaining to astrobiology, see Ćirković (2012).
72. Davies (1978); Tipler (1982). A similar argument in favor of finite cosmological past can be found in Epicurus and Lucretius (1997), as elaborated by Ćirković (2004).
73. However, see Lem ([1971] 1993) for a rational construal of this option.

31 The CMB and the multiverse:
Limits of scientific realism? (pp. 164–167)

74. See also the elaboration in Kragh (1996, 233–236).
75. Strictly speaking, if there are no event horizons (as in the EdS model), some of our copies will eventually enter our particle horizon at some late epoch. This statement, while technically true, represents a kind of an extreme stretch of the usual meaning of the words, since such a late epoch would be something like 10^{100} Hubble times, far beyond the effective heat death of the universe and the decay of all bound structures, including baryons themselves. It is an interesting general epistemological question whether such a remote prospect of empirical testing counts as scientific.
76. A different kind of multiverse appears in string theory/M-theory under names like string theory landscape, anthropic landscape (of string theory), or string multiverse. While this concept is clearly beyond the scope of the present book, it is important to mention it as a controversial facet of modern cosmology and its complex interaction with fundamental physics; for more details, see Bousso and Polchinski (2004); Susskind (2006). It has also been suggested that the CMB can serve as a testing ground for new, string-related physics (Hannestad & Mersini-Houghton, 2005).
77. Anthony Aguirre is the same researcher whom we mentioned above in relation to the Cold Big Bang models.
78. Peebles (2022).
79. This may not be surprising; the Hot Big Bang pioneers opted for an uneasy connection between initial conditions of the creation of the universe and the physical laws, as discussed in Chapter 30.

Appendix A Relativistic cosmological models (p. 169)

80. In general, cosmological density or cosmological density fraction of constituent X is defined as the density of X in units of critical density: $\Omega_x \equiv \rho_x / \rho_{\text{crit}} = \dfrac{8\pi G}{3H_0^2}\rho_x$.

Appendix B Dipole anisotropy (p. 174)

81. See also Humphreys, Maartens and Matravers (1997).

References

Abbasi, A., Hossain, L., Uddin, S., & Rasmussen, K. J. (2011). Evolutionary dynamics of scientific collaboration networks: Multi-levels and cross-time analysis. *Scientometrics* **89**(2), 687–710.

Acuña, P. (2021). Charting the landscape of interpretation, theory rivalry, and underdetermination in quantum mechanics. *Synthese*, **198**, 1711–1740.

Adams, P. J. (1983). Large numbers hypothesis. II: Electromagnetic radiation. *International Journal of Theoretical Physics* **22**, 421–436.

Ade, P. A. R., Rowan-Robinson, M., & Clegg, P. E. (1976). Millimetre emission from extragalactic objects. II Luminosities, spectra and contribution to the micro-wave background. *Astronomy & Astrophysics* **53**, 403–409.

Aguirre, A. (1999). Cold Big Bang nucleogenesis. *The Astrophysical Journal* **521**, 17–29.

Aguirre, A. (2000). The cosmic background radiation in a cold Big Bang. *Astrophysical Journal* **533**, 1–18.

Aguirre, A., & Johnson, M. C. (2011). A status report on the observability of cosmic bubble collisions. *Reports on Progress in Physics* **74**, 074901.

Albrow, M. G. (1973). CPT conservation in the oscillating model of the universe. *Nature Physical Science* **241**, 56–57.

Alexanian, M. (1970). Possible nonequilibrium nature of cosmic-background radiation. *Astrophysical Journal* **159**, 745–752.

Alfvén, H. O. (1979). Hubble expansion in a Euclidean framework. *Astrophysics and Space Science* **66**, 23–37.

Alfvén, H. O. (1984). Cosmology: Myth or science? *Journal of Astrophysics and Astronomy* **5**(1), 79–98.

Alfvén, H. O. (1990). Cosmology in the plasma universe: An introductory exposition. *IEEE Transactions on Plasma Science* **18**, 5–10.

Alfvén, H. O., & Mendis, A. (1977). Interpretation of observed cosmic microwave background radiation. *Nature* **266**, 698–699.

Alpher, R. A. (1962). Laboratory test of the Finlay-Freundlich red shift hypothesis. *Nature* **196**, 367–368.

Alpher, R. A., & Herman, R. C. (1948a). Evolution of the universe. *Nature* **162**, 774–775.

Alpher, R. A., & Herman, R. C. (1948b). On the relative abundance of the elements. *Physical Review* **74**, 1737–1742.

Alpher, R. A., & Herman, R. C. (1949). Remarks on the evolution of the expanding universe. *Physical Review* **75**, 1089–1095.

Alpher, R. A., Bethe, H., & Gamow, G. (1948). The origin of chemical elements. *Physical Review* **73**, 803–804.

Alvargonzález, D. (2013). Is the history of science essentially Whiggish? *History of Science*, **51**(1), 85–99.

Arp, H. C. (1987). *Quasars, Redshifts and Controversies*. Cambridge: Cambridge University Press.

Arp, H. C., & Van Flandern, T. (1992). The case against the big bang. *Physics Letters A* **164**, 263–273.

Arp, H. C., Burbidge, G., Hoyle, F., Narlikar, J. V., & Wickramasinghe, N. C. (1990). The extragalactic universe: An alternative view. *Nature* **346**, 807–812.

Barnes, L. A. (2012). The fine-tuning of the universe for intelligent life. *Publications of the Astronomical Society of Australia* **29**, 529–564.

Barrow, J. D. (1978). Quiescent cosmology. *Nature* **272**, 211–215.

Barrow, J. D., & Tipler, F. J. (1986). *The Anthropic Cosmological Principle*. New York: Oxford University Press.

Baryshev, Y. V., Raikov, A. A., & Tron, A. A. (1996). Microwave background radiation and cosmological large numbers. *Astronomical and Astrophysical Transactions* **10**, 135–138.

Beane, S. R., Davoudi, Z., & Savage, M. J. (2014). Constraints on the universe as a numerical simulation. *The European Physical Journal A* **50**, 148.

Beisbart, C. (2009). Can we justifiably assume the cosmological principle in order to break model underdetermination in cosmology? *Journal for General Philosophy of Science* **40**, 175–205.

Belousek, D. W. (2005). Underdetermination, realism, and theory appraisal: An epistemological reflection on quantum mechanics. *Foundations of Physics*, **35**, 669–695.

Boltzmann, L. (1895). On certain questions of the theory of gases. *Nature* **51**, 413–414.

Bond, J. R., Carr, B. J., & Hogan, C. J. (1991). Cosmic backgrounds from primeval dust. *Astrophysical Journal* **367**, 420–454.

Bond, J. R., Kofman, L., & Pogosyan, D. (1996). How filaments of galaxies are woven into the cosmic web. *Nature* **380**, 603–606.

Bondi, H. (1955). Fact and inference in theory and in observation. *Vistas in Astronomy* **1**, 155–162.

Bondi, H. (1960), *Cosmology*. Cambridge: Cambridge University Press.

Bondi, H. (1960). Gravitational waves in general relativity. *Nature* **186**(4724), 535–535.

Bondi, H. (1992). The philosopher for science. *Nature* **358**, 363.

Bondi, H., & Gold, T. (1948). The steady-state theory of the expanding universe. *Monthly Notices of the Royal Astronomical Society* **108**, 252–270.

Bostrom, N. (2002). *Anthropic Bias: Observation Selection Effects in Science and Philosophy*. New York: Routledge.

Bousso, R., & Polchinski, J. (2004). The string theory landscape. *Scientific American* **291**, 60–69.

Boyd, N. M. (2018). Evidence enriched. *Philosophy of Science* **85**(3), 403–421.

Boyd, N. M., & Matthiessen, D. (2023). Observations, experiments, and arguments for epistemic superiority in scientific methodology. *Philosophy of Science* **91**(1), 111–131.

Butterfield, H. [1931] (1959). *The Whig Interpretation of History*. London: G. Bell and Sons.

Butterfield, J. (2012). Underdetermination in cosmology: An invitation. *Aristotelian Society Supplementary Volume* **86**(1), 1–18. https://doi.org/10.1111/j.1467-8349.2012.00205.x

Butterfield, J. (2014). On under-determination in cosmology. *Studies in History and Philosophy of Science Part B: Studies in History and Philosophy of Modern Physics* **46**, 57–69.

Canuto, V., & Hsieh, S. H. (1977). Dirac cosmology and the microwave background. *Astronomy and Astrophysics* **61**, L5–L6.

Canuto, V., & Lodenquai, J. (1977). Dirac cosmology. *Astrophysical Journal* **211**, 342–356.

Canuto, V., Adams, P. J., Hsieh, S. H., & Tsiang, E. (1977). Scale-covariant theory of gravitation and astrophysical applications. *Physical Review D* **16**, 1643–1663.

Carr, B. (ed.) (2007). *Universe or Multiverse?* Cambridge: Cambridge University Press.

Carr, B. J. (1977). Black hole and galaxy formation in a cold early universe. *Monthly Notices of the Royal Astronomical Society (MNRAS)* **181**, 293–309.

Carr, B. J. (1981a). Pregalactic black hole accretion and the thermal history of the universe. *Monthly Notices of the Royal Astronomical Society* **194**, 639–668.

Carr, B. J. (1981b). Pregalactic stars and the origin of the microwave background. *Monthly Notices of the Royal Astronomical Society* **195**, 669–684.

Carr, B. J. (1994). Baryonic dark matter. *Annual Review of Astronomy and Astrophysics* **32**, 531–590.

Carr, B. J., & Rees, M. J. (1977). A tepid model for the early universe. *Astronomy & Astrophysics* 61, 705–709.

Carr, B. J., & Rees, M. J. (1979). The anthropic principle and the structure of the physical world. *Nature* **278**, 605–612.

Carrier, M. (2011). Underdetermination as an epistemological test tube: Expounding hidden values of the scientific community. *Synthese*, **180**(2), 189–204.

Carter, B. (1974). Large number coincidences and the anthropic principle in cosmology. In M. S. Longair (ed.), *Confrontation of Cosmological Theories with Observational Data; Proceedings of the Symposium, Krakow, Poland*, September 10–12, 1973 (pp. 291–298). Dordrecht: D. Reidel Publishing Co.

Chang, H. (2009). We have never been whiggish (About Phlogiston). *Centaurus* **51**, 239–264.

Chang, H. (2010). The hidden history of phlogiston. *HYLE – International Journal for Philosophy of Chemistry* **16**, 47–79.

Cheng, F. H. (1981). Estimation of the deceleration parameter q_0 using Quasar data. *Irish Astronomical Journal* **15**, 36–41.

Cheng, E. S., Saulson, P. R., Wilkinson, D. T., & Corey, B. E. (1979). Large-scale anisotropy in the 2.7 K radiation. *The Astrophysical Journal* **232**, L139–L143.

Ćirković, M. M. (2003). The thermodynamical arrow of time: Reinterpreting the Boltzmann–Schuetz argument. *Foundations of Physics* **33**, 467–490.

Ćirković, M. M. (2004). The anthropic principle and the duration of the cosmological past. *Astronomical & Astrophysical Transactions* **23**, 567–597.

Ćirković, M. M. (2012). *The Astrobiological Landscape: Philosophical Foundations of the Study of Cosmic Life*. Cambridge: Cambridge University Press.

Ćirković, M. M. (2016). Anthropic arguments outside of cosmology and string theory. *Belgrade Philosophical Annual* **29**, 91–114.

Ćirković, M. M., & Balbi, A. (2020). Copernicanism and the typicality in time. *International Journal of Astrobiology* **19**, 101–109.

Ćirković, M. M., & Perović, S. (2018). Alternative explanations of the cosmic microwave background: A historical and an epistemological perspective. *Studies in History and Philosophy of Science Part B: Studies in History and Philosophy of Modern Physics* **62**, 1–18.

Cleland, C. E. (2002). Methodological and epistemic differences between historical science and experimental science. *Philosophy of Science* **69**(3), 474–496.

Cleland, C. E. (2011). Prediction and explanation in historical natural science. *The British Journal for the Philosophy of Science* **62**, 551–582.

Coles, P., & Lucchin, F. (1995). *Cosmology: The Origin and Evolution of Cosmic Structure*. New York: John Wiley & Sons.

Conklin, E. K. (1969). Velocity of the Earth with respect to the cosmic background radiation. *Nature* **222**, 971–972.

Copi, C. J., Schramm, D. N., & Turner, M. S. (1995). Assessing big-bang nucleosynthesis. *Physics Review Letters* **75**, 3981–3984.

Cowan, J. J., Sneden, C., Burles, S., Ivans, I. I., Beers, T. C., Truran, J. W., Lawler, J. E., Primas, F., Fuller, G. M., Pfeiffer, B., & Kratz, K. L. (2002). The chemical composition and age of the metal-poor halo star BD+ 17 3248. *The Astrophysical Journal* **572**, 861–879.

Crawford, D. F. (1987a). Diffuse background X rays and the density of the intergalactic medium. *Australian Journal of Physics* **40**(3), 459–464.

Crawford, D. F. (1987b). Photons in curved space? Time. *Australian Journal of Physics* **40**(3), 449–458.

Crawford, D. F. (1991). A new gravitational interaction of cosmological importance. *Astrophysical Journal* **377**, 1–6.

Crawford, T. A., Hogg, D. C., & Hunt, L. E. (1961). A horn-reflector antenna for space communication. *Bell System Technical Journal* **40**(4), 1095–1116.

Crill, B. P., et al. (2003). BOOMERANG: A balloon-borne millimeter wave telescope and total power receiver for mapping anisotropy in the cosmic microwave background. *Astrophysical Journal Supplement Series* **148**, 527–541.

Currie, A. (2013). Convergence as evidence. *The British Journal for the Philosophy of Science* **64**(4), 763–786.

Currie, A., & Levy, A. (2019). Why experiments matter. *Inquiry* **62**(9–10), 1066–1090.

Cushing, J. T. (1994). *Quantum Mechanics: Historical Contingency and the Copenhagen Hegemony*. Chicago: University of Chicago Press.

da Silveira Ferreira, P., & Quartin, M. (2021). First constraints on the intrinsic CMB dipole and our velocity with Doppler and aberration. *Physical Review Letters* **127**, 101301.

Damour, T., Gibbons, G. W., & Taylor, J. H. (1988). Limits on the variability of G using binary-pulsar data. *Physical Review Letters* **61**, 1151–1154.

Davies, P. C. W. (1972). Closed time as an explanation of the black body background radiation. *Nature Physical Science* **240**, 3–5.

Davies, P. C. W. (1977). *The Physics of Time Asymmetry*. Berkeley: University of California Press.

Davies, P. C. W. (1978). Cosmic heresy? *Nature* **273**, 336–337.

Davis, M., & Peebles, P. J. E. (1983). A survey of galaxy redshifts. V. The two-point position and velocity correlations. *The Astrophysical Journal* **267**, 465–482.

Dawid, R. (2006). Underdetermination and theory succession from the perspective of string theory. *Philosophy of Science* **73**, 298–322.

Dawid, R., Hartmann, S., & Sprenger, J. (2015). The no alternatives argument. *The British Journal for the Philosophy of Science* **66**, 213–234.

Dicke, R. H. (1961). Dirac's Cosmology and Mach's Principle. *Nature* **192**, 440–441.

Dicke, R. H. (1962). Long-range scalar interaction. *Physical Review* **126**, 1875–1877.

Dicke, R. H., Beringer, R., Kyhl, R. L., & Vane, A. B. (1946). Atmospheric absorption measurements with a microwave radiometer. *Physical Review* **70**, 340–348.

Dicke, R. H., Peebles, P. J. E., Roll, P. G., & Wilkinson, D. T. (1965). Cosmic black-body radiation. *Astrophysical Journal* **142**, 414–419.

Dingle, H. (1953). Address. *Monthly Notices of the Royal Astronomical Society* **113**, 398.

Dingle, H. (1954). Science and modern cosmology. *Science* **120**, 513–521.

Dirac, P. A. M. (1974). Cosmological models and the large numbers hypothesis. *Proceedings of the Royal Society of London A* **338**, 439–446.

Dirac, P. A. M. (1979). The large numbers hypothesis and the Einstein Theory of Gravitation. *Proceedings of the Royal Society of London A* **365**, 19–30.

Disney, M. J. (2000). The case against cosmology. *General Relativity and Gravitation* **32**, 1125–1134.

Dodelson, S. (2003). *Modern Cosmology*. London: Academic Press.

Doroshkevich, A. G., & Novikov, I. D. (1964). Mean density of radiation in the meta-galaxy and certain problems in relativistic cosmology. *Soviet Physics Doklady* **9**, 4292–4298.

Dressler, A., Lynden-Bell, D., Burstein, D., Davies, R. L., Faber, S. M., Terlevich, R., & Wegner, G. (1987). Spectroscopy and photometry of elliptical galaxies. I-A new distance estimator. *The Astrophysical Journal* **313**, 42–58.

Duhem, P. M. M. (1994). *The Aim and Structure of Physical Theory* (Vol. 13). Originally published in 1914 as *La Théorie Physique: Son Objet et sa Structure* (Paris: Marcel Riviera & Cie.). Princeton: Princeton University Press.

Earman, J. (1987). The SAP also rises: A critical examination of the anthropic principle. *American Philosophical Quarterly* **24**, 307–317.

Earman, J. (1993). Underdetermination, realism and reason. *Midwest Studies in Philosophy* **18**, 19–38.

Egan, C. A., & Lineweaver, C. H. (2010). A larger estimate of the entropy of the universe. *The Astrophysical Journal* **710**, 1825–1834.

Eichler, D. (1977). Primeval entropy fluctuations and the present-day pattern of gravitational clustering. *Astrophysical Journal* **218**, 579–581.

Einasto, J., Kaasik, A., & Saar, E. (1974). Dynamic evidence on massive coronas of galaxies. *Nature* **250**, 309–310.

Einasto, M., Lietzen, H., Gramann, M., Tempel, E., Saar, E., Liivamägi, L. J., Heinämäki, P., Nurmi, P., & Einasto, J. (2016). Sloan Great Wall as a complex of superclusters with collapsing cores. *Astronomy & Astrophysics* **595**, A70.

Einstein, A. (1917). Kosmologische Betrachtungen zur allgemeinen Relativitätstheorie, Sitzungsberichte der Königlich Preußischen Akademie der Wissenschaften (Berlin) 142–152.

Ellis, G. F. R. (1978). Is the universe expanding? *General Relativity and Gravitation* **9**, 87–94.

Ellis, G. F. R. (1984). Alternatives to the Big Bang. *Annual Review of Astronomy and Astrophysics* **22**, 157–184.

Ellis, G. F. R. (2014). On the philosophy of cosmology. *Studies in History and Philosophy of Science Part B: Studies in History and Philosophy of Modern Physics* **46**, 5–23.

Ellis, G. F. R., & Brundrit, G. B. (1979). Life in the infinite universe. *Quarterly Journal of the Royal Astronomical Society* **20**, 37–41.

Ellis, G. F. R., Kirchner, U., & Stoeger, W. R. (2004). Multiverses and physical cosmology. *Monthly Notices of the Royal Astronomical Society* **347**, 921–936.

Ellis, G. F. R., Maartens, R., & Nel, S. D. (1978). The expansion of the Universe. *Monthly Notices of the Royal Astronomical Society* **184**, 439–465.

Engelhardt, H. T., & Caplan, A. L. (1987). *Scientific Controversies: Case studies in the Resolution and Closure of Disputes in Science and Technology*. Cambridge: Cambridge University Press.

Fabbri, R., Guidi, I., Melchiorri, F., & Natale, V. (1980). Measurement of the cosmic-background large-scale anisotropy in the millimetric region. *Physical Review Letters* **44**, 1563–1566.

Fabian, A. C., & Barcons, X. (1992). The origin of the X-ray background. *Annual Review of Astronomy and Astrophysics* **30**, 429–456.

Fahr, H. J., & Zoennchen, J. H. (2009). The "writing on the cosmic wall": Is there a straightforward explanation of the cosmic microwave background? *Annals of Physics*, **18**, 699–721.

Feeney, S. M., Johnson, M. C., Mortlock, D. J., & Peiris, H. V. (2011). First observational tests of eternal inflation. *Physical Review Letters* **107**, 071301.

Fixsen, D. J. (2009). The temperature of the cosmic microwave background. *The Astrophysical Journal* **707**, 916–920.

Fixsen, D. J., Cheng, E. S., Gales, J. M., Mather, J. C., Shafer, R. A., & Wright, E. L. (1996). The cosmic microwave background spectrum from the full COBE FIRAS data set. *Astrophysical Journal* **473**, 576–587.

Foley, R. J., Filippenko, A. V., Leonard, D. C., Riess, A. G., Nugent, P., & Perlmutter, S. (2005). A definitive measurement of time dilation in the spectral evolution of the moderate-redshift type Ia Supernova 1997ex. *The Astrophysical Journal* **626**, L11–L14.

Fortey, R. (2005). *Earth: An Intimate History*. New York: Vintage Books.

Fraser, D. (2009). Quantum field theory: Underdetermination, inconsistency, and idealization. *Philosophy of Science*, **76**(4), 536–567.

Frautschi, S. (1982). Entropy in an expanding universe. *Science* **217**, 593–599.

Friedmann, A. [1922] (1979). On the curvature of space. In Kenneth R. Lang, & Owen Gingerich (eds.), *A Source Book in Astronomy and Astrophysics, 1900–1975* (pp. 838–843). Cambridge: Harvard University Press.

Frigg, R., & Hartmann, S. (2018). Stanford encyclopedia of philosophy: Models in science. *Stanford Encyclopedia of Philosophy*, https://plato.stanford.edu/archives/spr2020/entries/models-science/.

Gallagher, S., & Smeenk C. (2023). What's in a survey? Simulation-induced selection effects in astronomy. In N. M. Boyd, S. De Baerdemaeker, K. Heng, & V. Matarese (eds.), *Philosophy of Astrophysics: Stars, Simulations, and the Struggle to Determine What Is Out There* (pp. 207–222). Springer.

Gamow, G. (1946). Expanding universe and the origin of elements. *Physical Review* **70**, 572–573.

Gamow, G. (1948). The origin of elements and the separation of galaxies. *Physical Review* **74**, 505–506.

Gamow, G. (1949). On relativistic cosmogony. *Reviews of Modern Physics* **21**, 367–373.

Gee, H. (1999). *In Search of Deep Time: Beyond the Fossil Record to a New History of Life*. Ithaca, NY: Cornell University Press.

Giunta, C. J. (2022). Is There Room for the Present in the History of Science? *Bulletin for the History of Chemistry*, **47**(1), 163–170.

Gnedin, N. Y. (2000). Effect of reionization on structure formation in the universe. *The Astrophysical Journal* **542**, 535–541.

Gnedin, N. Y., & Ostriker, J. P. (1992). Light element nucleosynthesis – A false clue? *Astrophysical Journal* **400**, 1–20.

Gold, T., & Pacini, F. (1968). Can the observed microwave background BE due to a super-position of sources? *Astrophysical Journal* **152**, L115–L118.

Gorenstein, M. V., & Smoot, G. F. (1981). Large-angular-scale anisotropy in the cosmic background radiation. *Astrophysical Journal* **244**, 361–381.

Gould, S. J. (2002). *The Structure of Evolutionary Theory*. Cambridge: Belknap Press.

Gould, R. J., & Ramsay, W. (1966). The temperature of intergalactic matter. *Astrophysical Journal* **144**, 587.

Gould, S. J., & Vrba, E. S. (1982). Exaptation: A missing term in the science of form. *Paleobiology* **8**, 4–15.

Gregory, J. (2005). *Fred Hoyle's Universe*. Oxford: Oxford University Press.

Gush, H. P. (1981). Rocket measurement of the cosmic background submillimeter spectrum. *Physical Review Letters* **47**(10), 745.

Hall, A. R. (1983). On whiggism. *History of Science* **21**, 45–59.

Halpern, P. (2021). *Flashes of Creation: George Gamow, Fred Hoyle, and the Great Big Bang Debate*. New York: Basic Books.

Hannestad, S., & Mersini-Houghton, L. (2005). First glimpse of string theory in the sky? *Physical Review D* **71**, 123504.

Harrison, E. (1987). Whigs, prigs and historians of science. *Nature* **329**, 213–214.

Harwit, M. (2019). *Cosmic Discovery: The Search, Scope, and Heritage of Astronomy*. Cambridge: Cambridge University Press.

Hayakawa, S. (1984). Cosmic background radiation from pregalactic objects. *Advances in Space Research* **3**, 449–457.

Hazard, C., & Salpeter, E. E. (1969). Discrete sources and the microwave background in steady-state cosmologies. *Astrophysical Journal* **157**, L87–L90.

Hellings, R. W. et al. (1983). Experimental test of the variability of G using Viking lander ranging data. *Physical Review Letters* **51**, 1609–1612.

Hinshaw, G. et al. (2003). First-year Wilkinson Microwave Anisotropy Probe (WMAP) observations: The angular power spectrum. *Astrophysical Journal Supplement Series* **148**, 135–159.

Hinshaw, G. et al. (2013). Nine-year Wilkinson Microwave Anisotropy Probe (WMAP) observations: Cosmological parameter results. *Astrophysical Journal Supplement* **208**, 25article id. 19.

Hogarth, J. E. (1962). Cosmological considerations of the absorber theory of radiation. *Proceedings of the Royal Society of London A* **267**, 365–383.

Holton, G. (1988). *Thematic Origins of Scientific Thought: Kepler to Einstein*. Cambridge: Harvard University Press.

Hossenfelder, S. (2018). *Lost in Math: How Beauty Leads Physics Astray*. New York: Hachette UK.

Hoyle, F. (1948). A new model for the expanding universe. *Monthly Notices of the Royal Astronomical Society (MNRAS)* **108**, 372–382.

Hoyle, F. (1955). *Frontiers of Astronomy*. New York: Harper and Row.

Hoyle, F. (1975). On the origin of the microwave background. *Astrophysical Journal* **196**, 661–670.

Hoyle, F. (1983). *The Intelligent Universe*. London: Michael Joseph Limited.

Hoyle, F. (1994). *Home Is Where the Wind Blows: Chapters from a Cosmologist's Life*. Mill Valley: University Science Books.

Hoyle, F., & Burbidge, G. (1992). Possible explanations of the large angle fluctuations of the microwave background. *Astrophysical Journal, Part 2-Letters* **399**, 1, L9–L10.

Hoyle, F., & Narlikar, J. V. (1964). Time symmetric electrodynamics and the arrow of time in cosmology. *Proceedings of the Royal Society of London A* **277**, 1–23.

Hoyle, F., & Narlikar, J. V. (1966). A radical departure from the 'steady-state' concept in cosmology. *Proceedings of the Royal Society of London A* **290**, 162–176.

Hoyle, F., & Narlikar, J. V. (1971). Electrodynamics of direct interparticle action. II: Relativistic treatment of radiative processes. *Annals of Physics* **62**, 44–97.

Hoyle, F., & Narlikar, J. V. (1972a). Cosmological models in a conformally invariant gravitational theory-I. The Friedmann models. *Monthly Notices of the Royal Astronomical Society* **155**, 305–321.

Hoyle, F., & Narlikar, J. V. (1972b). Cosmological models in a conformally invariant gravitational theory-II. A new model. *Monthly Notices of the Royal Astronomical Society* **155**, 323–335.

Hoyle, F., & Sandage, A. (1956). The second-order term in the redshift-magnitude relation. *Publications of the Astronomical Society of the Pacific* **68**(403), 301–307.

Hoyle, F., & Wickramasinghe, N. C. (1967). Impurities in interstellar grains. *Nature* **214**, 969–971.

Hoyle, F., Burbidge, G. R., & Narlikar, J. V. (1993). A quasi-steady state cosmological model with creation of matter. *Astrophysical Journal* **410**, 437–457.

Hoyle, F., Burbidge, G. R., & Narlikar, J. V. (1994). Astrophysical deductions from the quasi-steady-state cosmology. *Monthly Notices of the Royal Astronomical Society* **267**, 1007–1019.

Hoyle, F., Burbidge, G., & Narlikar, J. V. (1999). *A Different Approach to Cosmology: From a Static Universe through the Big Bang towards Reality*. Cambridge: Cambridge University Press.

Hoyle, F., Dunbar, D. N. F., Wenzel, W. A., & Whaling, W. (1953). A state in C-12 predicted from astrophysical evidence. *Physical Review* **92**, 1095–1095.

Hsu, S., & Zee, A. (2006). Message in the sky. *Modern Physics Letters A* **21**, 1495–1500.

Humphreys, N. P., Maartens, R., & Matravers, D. R. (1997). Anisotropic observations in universes with nonlinear inhomogeneity. *The Astrophysical Journal* **477**, 47–57.

Ijjas, A., Steinhardt, P. J., & Loeb, A. (2014). Inflationary schism. *Physics Letters* **B736**, 142–146.

Islam, J. N. (2004). *An Introduction to Mathematical Cosmology* (2nd ed.). Cambridge: Cambridge University Press.

Jardine, N. (2003). Whigs and stories: Herbert Butterfield and the historiography of science. *History of Science* **41**, 125–140.

Kant, I. [1797] (2005). Universal natural history and theory of heaven. *Oxford Text Archive Core Collection*.

Klee, R. (2002). The revenge of Pythagoras: How a mathematical sharp practice undermines the contemporary design argument in Astrophysical Cosmology. *British Journal for the Philosophy of Science* **53**, 331–354.

Kolb, E. W., & Turner, M. S. (1990). *The Early Universe*. Boulder: Westview Press.

Komatsu, E. et al. (2011). Seven-Year Wilkinson Microwave Anisotropy Probe (WMAP) observations: Cosmological interpretation. *The Astrophysical Journal Supplement Series* **192**, 18 (47pp).

Kragh, H. (1982). Cosmo-physics in the thirties: Towards a history of Dirac cosmology. *Historical Studies in the Physical Sciences* **13**, 69–108.

Kragh, H. (1996). *Cosmology and Controversy*. Princeton: Princeton University Press.

Kragh, H. (1997). Remarks on the historiography and philosophy of modern cosmology. *Danish Yearbook of Philosophy* **32**, 65–86.

Kragh, H. (2004). *Matter and Spirit in the Universe: Scientific and Religious Preludes to Modern Cosmology*. London: Imperial College Press.

Kragh, H. (2011). *Higher Speculations: Grand Theories and Failed Revolutions in Physics and Cosmology*. Oxford: Oxford University Press.

Kragh, H. (2012). Quasi-steady-state and related cosmological models: A historical review. *arXiv:1201.3449* **[physics.hist-ph]**.

Kragh, H. (2013a). The most philosophically important of all the sciences: Karl Popper and physical cosmology. *Perspectives on Science* **21**, 325–357.

Kragh, H. (2013b). Cyclic models of the relativistic universe: The early history. *arXiv:1308.0932*.

Kragh, H. (2015). *Masters of the Universe: Conversations with Cosmologists of the Past*. Oxford: Oxford University Press.

Kragh, H. (2016). *Varying Gravity: Dirac's Legacy in Cosmology and Geophysics*. Basel: Birkhäuser Verlag.

Kragh, H., & Longair, M. S. (2019). *The Oxford Handbook of the History of Modern Cosmology*. Oxford: Oxford University Press.

Kuhn, T. (1962). *The Structure of Scientific Revolutions*. Chicago: Chicago University Press.

Lahav, O., Kaiser, N., & Hoffman, Y. (1990). Local gravity and peculiar velocity-Probes of cosmological models. *The Astrophysical Journal* **352**, 448–456.

Lakatos, I. (1978). *The Methodology of Scientific Research Programmes, 1*. Cambridge: Cambridge University Press.

LaViolette, P. A. (1986). Is the universe really expanding? *Astrophysical Journal* **301**, 544–553.

Layzer, D. (1968). Black-body radiation in a cold universe. *Astrophysical Letters* **1**, 99–102.

Layzer, D. (1972). Science or superstition? (A physical scientist looks at the IQ controversy). *Cognition* **1**, 265–299.

Layzer, D. (1976). The arrow of time. *Astrophysical Journal* **206**, 559–569.

Layzer, D. (1992). On the origin of cosmic structure. *Astrophysical Journal* **392**, L5–L8.

Layzer, D., & Hively, R. (1973). Origin of the microwave background. *Astrophysical Journal* **179**, 361–370.

Lem, S. [1971] (1993). The new cosmogony. In *A Perfect Vacuum* translated by M. Kandel (pp. 197–227). Evanston: Northwestern University Press.

Lemaître, G. (1931). The beginning of the world from the point of view of quantum theory. *Nature* **127**(3210), 706.

Lerner, E. J. (1988). Plasma model of microwave background and primordial elements – an alternative to the Big Bang. *Laser and Particle Beams* **6**, 457–469.

Lerner, E. J. (1991). *The Big Bang Never Happened*. New York: Times Books.

Lerner, E. J. (1995). Intergalactic radio absorption and the COBE data. *Astrophysics and Space Science* **227**, 61–81.

Li, A. (2003). Cosmic needles versus cosmic microwave background radiation. *The Astrophysical Journal* **584**, 593–598.

Lightman, A. P., & Rybicki, G. B. (1979). Inverse Compton reflection-Time-dependent theory. *Astrophysical Journal* **232**, 882–890.

Linde, A. D. (1983). Chaotic inflation. *Physics Letters B*, **129**(3–4), 177–181.

Linde, A. D. (1990). *Inflation and Quantum Cosmology*. Boston: Academic Press.

Linde, A. D. (2008). Inflationary cosmology. In M. Lemoine, J. Martin, & P. Peters (eds.), *Inflationary Cosmology* (pp. 1–51). Berlin: Springer.

Linde, A. D. (2014). Inflationary Cosmology after Planck 2013, Les Houches School 2013 Lectures (preprint https://arxiv.org/abs/1402.0526).

Longair, M. S. (1993). Modern cosmology: A critical assessment. *Quarterly Journal of the Royal Astronomical Society* **34**, 157–199.

López-Corredoira, M. (2013). Peaks in the CMBR power spectrum II: Physical interpretation for any cosmological scenario. *International Journal of Modern Physics D* **22**, 1350032.

López-Corredoira, M. (2014). Non-standard models and the sociology of cosmology. *Studies in History and Philosophy of Modern Physics* **46A**, 86–96.

Lucretius (1997). *On the Nature of Things*. Amherst: Prometheus Books.

Magueijo, J., & Land, K. (2006). Template fitting and the large-angle cosmic microwave background anomalies. *Monthly Notices of the Royal Astronomical Society* **367**(4), 1714–1720.

Maher, P. (1993). Discussion: Howson and Franklin on Prediction. *Philosophy of Science*, **60**(2), 329–340.

Mansfield, V. N. (1976). Dirac cosmologies and the microwave background. *Astrophysical Journal* **210**, L137–L138.

Mather, J., & Boslough, J. (1997). *The Very First Light: The True Inside Story of the Scientific Journey Back to the Dawn of the Universe*. New York: Basic Books.

Mather, J. C., Cheng, E. S., Eplee Jr, R. E., Isaacman, R. B., Meyer, S. S., Shafer, R. A., et al. (1990). A preliminary measurement of the cosmic microwave background spectrum by the Cosmic Background Explorer (COBE) satellite. *The Astrophysical Journal* **354**, L37–L40.

Mather, J. C. et al. (1994). Measurement of the cosmic microwave background spectrum by the COBE FIRAS instrument. *Astrophysical Journal* **420**, 439–444.

Matsubara, K. (2013). Realism, underdetermination and string theory dualities. *Synthese*, **190**(3), 471–489.

Matsumoto, T., Hayakawa, S., Matsuo, H., Murakami, H., Sato, S., Lange, A. E., & Richards, P. L. (1988). The submillimeter spectrum of the cosmic background radiation. *The Astrophysical Journal* **329**, 567–571.

Mayr, E. (1990). When is historiography whiggish? *Journal of the History of Ideas* **51**, 301–309.

McCrea, W. H. (1953). Cosmology. *Reports on Progress in Physics* **16**, 321–363.

McKellar, A. (1941). Molecular lines from the lowest states of diatomic molecules composed of atoms probably present in interstellar space. *Publications of the Dominion Astrophysical Observatory* **7**, 251–272.

Mercier, C. (2019). Calculation of the universal gravitational constant, of the hubble constant, and of the average CMB temperature. *Journal of Modern Physics* **10**, 641–662.

Millano, A. D., Jusufi, K., & Leon, G. (2023). Phase space analysis of the bouncing universe with stringy effects. *Physics Letters B* **841**, 137916.

Milojević, S. (2014). Principles of scientific research team formation and evolution. *Proceedings of the National Academy of Sciences* **111**(11), 3984–3989.

Misner, C. W. (1968). The isotropy of the universe. *Astrophysical Journal* **151**, 431–457.

Misner, C. W., Thorne, K. S., & Wheeler, J. A. (1973). *Gravitation*. San Francisco: W. H. Freeman and Co.

Mitton, S. (2011). *Fred Hoyle: A Life in Science*. Cambridge: Cambridge University Press.

Molaro, P., Levshakov, S. A., Dessauges-Zavadsky, M., & D'Odorico, S. (2002). The cosmic microwave background radiation temperature at toward QSO 0347–3819. *Astronomy & Astrophysics* **381**, L64–L67.

Moradpour, H., Shabani, H., Ziaie, A. H., & Sharma, U. K. (2021). Non-minimal coupling inspires the Dirac cosmological model. *The European Physical Journal Plus* **136**, 1–12.

Mukhopadhyay, U., & Ray, S. (2014). N.C. Rana: The Life of a "Comet" in the Astrophysical World, preprint https://arxiv.org/abs/1401.2141v1.

Nadolny, T., Durrer, R., Kunz, M., & Padmanabhan, H. (2021). A new way to test the Cosmological Principle: Measuring our peculiar velocity and the large-scale anisotropy independently. *Journal of Cosmology and Astroparticle Physics* **11**, 009.

Narlikar, J. V. (1983). *Introduction to Cosmology*. Boston: Jones and Bartlett Publishers.

Narlikar, J. V., & Rana, N. C. (1980). Cosmic microwave background spectrum and G-varying cosmology. *Physics Letters A* **77**, 219–220.

Narlikar, J. V., & Rana, N. C. (1983). Cosmic microwave background spectrum in the Hoyle-Narlikar cosmology. *Physics Letters A* **99**, 75–76.

Narlikar, J. V., & Wickramasinghe, N. C. (1968). Interpretation of cosmic microwave background. *Nature* **217**, 1235–1236.

Narlikar, J. V., Vishwakarma, R. G., Hajian, A., Souradeep, T., Burbidge, G., & Hoyle, F. (2003). Inhomogeneities in the microwave background radiation interpreted within the framework of the quasi-steady state cosmology. *The Astrophysical Journal* **585**, 1–11.

Newton-Smith, W. (1978). The underdetermination of theory by data. In R. Hilpinen *Rationality in Science: Studies in the Foundations of Science and Ethics* (pp. 91–110). Dordrecht: Springer Netherlands.

North, J. (1994). *The Fontana History of Astronomy and Cosmology*. London: Fontana Press.

Norton, J. D. (2008). The dome: An unexpectedly simple failure of determinism. *Philosophy of Science*, **75**(5), 786–798.

Norton, J. D. (2017). "Inference to the best explanation: Examples" draft chapter for the Material theory of induction, www.pitt.edu/~jdnorton/homepage/cv.html.

Okasha, S. (2001). What did Hume really show about induction? *Philosophical Quarterly* **51**(204), 307–327.

O'Raifeartaigh, C., McCann, B., Nahm, W., & Mitton, S. (2014). Einstein's exploration of a steady-state model of the universe. *The European Physical Journal H*, **39**(3), 353–367.

Oreskes, N. (2013). Why I am a presentist. *Science in Context* **26**(4), 595–609.

Ostriker, J. P., Peebles, P. J. E., & Yahil, A. (1974). The size and mass of galaxies, and the mass of the universe. *Astrophysical Journal* **193**, L1–L4.

Paczynski, B., & Piran, T. (1990). A dipole moment of the microwave background as a cosmological effect. *The Astrophysical Journal* **364**, 341–348.

Page, S. E. (2011). *Diversity and Complexity*. Princeton: Princeton University Press.

Pagels, H. R. (1998). A Cozy cosmology. In J. Leslie (ed.), *Modern Cosmology & Philosophy* (pp. 180–186). Amherst: Prometheus Books.

Pais, A. (1986). *Inward Bound: Of Matter and Forces in the Physical World*. Oxford: Oxford University Press.

Partridge, R. B. (1995). *3K: The Cosmic Microwave Background Radiation*. Cambridge: Cambridge University Press.

Partridge, R. B., & Wilkinson, D. T. (1967). Isotropy and homogeneity of the universe from measurements of the cosmic microwave background. *Physical Review Letters* **18**, 557–559.

Paul, P., Sengupta, R., & Ray, S. (2023). Studies on modified power law inflation. *Chinese Physics C* **47**, 035107.

Peacock, J. A. (1999). *Cosmological Physics*. Cambridge: Cambridge University Press.

Peebles, P. J. E. (1993). *Principles of Physical Cosmology*. Princeton: Princeton University Press.

Peebles, P. J. E. (1999). Penzias & Wilson's discovery of the cosmic microwave background. *Astrophysical Journal*, Centennial Issue, **525C**, 1067–1068.

Peebles, P. J. E. (2014). Discovery of the Hot Big Bang: What happened in 1948. *The European Physical Journal H*, **39**, 205–223.

Peebles, P. J. E. (2020). *Cosmology's Century: An Inside History of Our Modern Understanding of the Universe*. Princeton: Princeton University Press.

Peebles, P. J. E. (2022a). Anomalies in physical cosmology. *Annals of Physics* **447**, 169159.

Peebles, P. J. E. (2022b). *The Whole Truth: A Cosmologist's Reflections on the Search for Objective Reality*. Princeton: Princeton University Press.

Peebles, P. J., & Yu, J. T. (1970). Primeval adiabatic perturbation in an expanding universe. *The Astrophysical Journal* **162**, 815–836.

Peebles, P. J. E., Page, L. A., Jr., & Partridge, R. B. (2009). *Finding the Big Bang*. Cambridge: Cambridge University Press.

Penrose, R. (1979). Singularities and time-asymmetry. In S. W. Hawking, & W. Israel (eds.), *General Relativity: An Einstein Centenary* (pp. 581–638). Cambridge: Cambridge University Press.

Penrose, R. (1989). *The Emperor's New Mind*. Oxford: Oxford University Press.

Penzias, A. A., & Wilson, R. W. (1965). A measurement of excess antenna temperature at 4080 Mc/s. *Astrophysical Journal* **142**, 419–421.

Perović, S. (2011). Missing experimental challenges to the standard model of particle physics. *Studies in History and Philosophy of Science Part B: Studies in History and Philosophy of Modern Physics* **42**(1), 32–42.

Perović, S. (2018). Egalitarian paradise or factory drudgery? Organizing knowledge production in high energy physics (HEP) laboratories. *Social Epistemology* **32**(4), 241–261.

Perović, S. (2021). Observation, experiment, and scientific practice. *International Studies in the Philosophy of Science* **34**(1), 1–20.

Perović, S., Radovanović, S., Sikimić, V., & Berber, A. (2016). Optimal research team composition: Data envelopment analysis of Fermilab experiments. *Scientometrics* **108**, 83–111.

Pfenniger, D., Combes, F., & Martinet, L. (1994). Is dark matter in spiral galaxies cold gas? I. Observational constraints and dynamical clues about galaxy evolution. *Astronomy and Astrophysics* **285**, 79–93.

Phillips, P. R. (1994a). Solution of the field equations for a steady-state cosmology in a closed space. *Monthly Notices of the Royal Astronomical Society* **269**, 771–778.

Phillips, P. R. (1994b). Development of the closed steady-state cosmological model. *Monthly Notices of the Royal Astronomical Society* **271**, 499–503.

Pietsch, W. (2011). The Underdetermination debate: How lack of history leads to bad philosophy. In R. Cohen, K. Gavroglu, J. Renn (eds.), *Integrating History and Philosophy of Science: Problems and Prospects* (pp. 83–106). Dordrecht: Springer Netherlands.

Popper, K. (1972). *Objective Knowledge: An Evolutionary Approach.* Oxford: Oxford University Press.

Popper, K. R. (1992). *Conjectures and Refutations: The Growth of Scientific Knowledge* (5th ed.). London: Routledge.

Price, H. (1991). The asymmetry of radiation: Reinterpreting the Wheeler-Feynman argument. *Foundations of Physics* **21**, 959–975.

Puget, J. L., & Heyvaerts, J. (1980). Population III stars and the shape of the cosmological black body radiation. *Astronomy & Astrophysics* **83**, L10–L12.

Quine, W. V. (1975). On empirically equivalent systems of the world. *Erkenntnis* 313–328.

Rana, N. C. (1979). Absorption effects of intergalactic natural graphite whiskers on observations at microwave and radio frequencies. *Astrophysics and Space Science* **66**, 173–190.

Rana, N. C. (1980). Absorption effects due to intergalactic long whiskers of pyrolytic graphite and the cosmic microwave background. *Astrophysics and Space Science* **71**, 123–133.

Rana, N. C. (1981). Cosmic thermalization and the microwave background radiation. *Monthly Notices of the Royal Astronomical Society* **197**, 1125–1137.

Rassat, A., Starck, J. L., Paykari, P., Sureau, F., & Bobin, J. (2014). Planck CMB anomalies: Astrophysical and cosmological secondary effects and the curse of masking. *Journal of Cosmology and Astroparticle Physics* **2014**(08), 006.

Raup, D. (1999). *Nemesis Affair Revised and Expanded: A Story of the Death of the Dinosaurs and the Ways of Science.* New York: W. W. Norton & Company.

Rees, M. J. (1972). Origin of the cosmic microwave background radiation in a chaotic universe. *Physical Review Letters* **28**, 1669–1671.

Rees, M. J. (1978). Origin of pregalactic microwave background. *Nature* **275**, 35–37.

Rees, M. J. (2007). Cosmology and the multiverse. In B. Carr (ed.), *Universe or Multiverse?* (pp. 57–76). Cambridge: Cambridge University Press.

Roth, K. C., Meyer, D. M., & Hawkins, I. (1993). Interstellar cyanogen and the temperature of the cosmic microwave background radiation. *Astrophysical Journal Letters* **413**, L67–L71.

Rowan-Robinson, M. (1974). A discrete source model of the microwave background. *Monthly Notices of the Royal Astronomical Society* **168**, 45–50.

Rowan-Robinson, M. (1977). On the unity of activity in galaxies. *Astrophysical Journal* **213**, 635–647.

Rowan-Robinson, M., Negroponte, J., & Silk, J. (1979). Distortions of the cosmic microwave background spectrum by dust. *Nature* **281**, 635–638.

Sachs, R. K., & Wolfe, A. M. (1967). Perturbations of a cosmological model and angular variations of the microwave background. *Astrophysical Journal* **147**, 73–90.

Sakharov, A. D. (1967). Violation of CP invariance, C asymmetry, and baryon asymmetry of the universe. *Journal of Experimental and Theoretical Physics Letters* **5**, 24–27.

Salem, M. P. (2012). The CMB and the measure of the multiverse. *Journal of High Energy Physics* **2012**(6), 1–28.

Sandage, A. (1988). Observational tests of world models. *Annual Review of Astronomy and Astrophysics* **26**, 561–630.

Schild, R. E., & Gibson, C. H. (2008). Goodness in the Axis of Evil. *arXiv preprint arXiv:0802.3229*.

Schneider, R., Ferrara, A., Salvaterra, R., Omukai, K., & Bromm, V. (2003). Low-mass relics of early star formation. *Nature* **422**, 869–871.

Schramm, D. N., Michael, S., & Turner, M. S. (1998). Big-bang nucleosynthesis enters the precision era. *Reviews of Modern Physics* **70**, 303–318.

Sciama, D. W. (1955). On the formation of galaxies in a steady state universe. *Monthly Notices of the Royal Astronomical Society* **115**(1), 3–14.

Sciama, D. W. (1966). On the origin of the microwave background radiation. *Nature* **211**, 277–279.

Seo, H. J., & Eisenstein, D. J. (2003). Probing dark energy with baryonic acoustic oscillations from future large galaxy redshift surveys. *The Astrophysical Journal* **598**, 720–740.

Sepkoski, D. (2020). *Catastrophic Thinking: Extinction and the Value of Diversity from Darwin to the Anthropocene*. Chicago: University of Chicago Press.

Šešelja, D. (2022). Agent-based models of scientific interaction. *Philosophy Compass* **17**(7), e12855.

Setti, G. (1970). Infrared background from Seyfert galaxies. *Nature* **227**, 586–587.

Shakeshaft, J. R., & Webster, A. S. (1968). Microwave background in a steady state universe. *Nature* **217**, 339.

Sharov, A. S., & Novikov, I. D. (1993). *Edwin Hubble: Discoverer of the Big Bang Universe*. New York: Cambridge University Press.

Shiryaev, A. N. (2016). *Probability-1*. New York: Springer.

Sklar, L. (1975). Methodological conservatism. *The Philosophical Review* **84**(3), 374–400.

Smeenk, C., & Ellis. G. E. (2017). Philosophy of Cosmology, Stanford Encyclopedia of Philosophy, as accessed at https://plato.stanford.edu/entries/cosmology/ (last accessed March 30, 2024).

Smith, M. G., & Partridge, R. B. (1970). Can discrete sources produce the cosmic microwave radiation? *Astrophysical Journal* **159**, 737.

Smolin, L. (1997). *The Life of the Cosmos*. Oxford: Oxford University Press.

Smolin, L. (2004). Scientific alternatives to the anthropic principle. preprint arXiv:hep-th/0407213.

Smolin, L. (2006). *The Trouble with Physics: The Rise of String Theory, the Fall of a Science, and What Comes Next*. New York: Houghton Mifflin Company.

Smoot, G. F., Gorenstein, M. V., & Muller, R. A. (1977). Detection of anisotropy in the cosmic blackbody radiation. *Physical Review Letters* **39**, 898–901.

Smoot, G. F., Bensadouin, M., Bersanelli, M., De Amici, G., Kogut, A., Levin, S., & Witebsky, C. (1987). Long-wavelength measurements of the cosmic microwave background radiation spectrum. *Astrophysical Journal* **317**, L45–L49.

Smoot, G. F. et al. (1992). Structure in the COBE differential microwave radiometer first-year maps. *Astrophysical Journal* **396**, L1–L5.

Soler, L. (2023). What would it be like to be Bohmians? Experiencing a Gestalt Switch in Physics as an Effect of Path Dependence. *Social Epistemology*, DOI: 10.1080/02691728.2023.2212372

Soler, L., Trizio, E., & Pickering, A. (eds.) (2016). *Science as It Could Have Been: Discussing the Contingency/Inevitability Problem.* Pittsburgh: University of Pittsburgh Press.

Sorrell, W. H. (2008). The cosmic microwave background radiation in a non-expanding universe. *Astrophysics and Space Science* **317**, 59–70.

Spergel, D. N. et al. (2007). Three-year Wilkinson Microwave Anisotropy Probe (WMAP) observations: Implications for cosmology. *Astrophysical Journal Supplement Series* **170**, 377–408.

Springel, V., White, S. D., Jenkins, A., Frenk, C. S., Yoshida, N., Gao, L., Navarro, J., Thacker, R., Croton, D., Helly, J., & Peacock, J. A. (2005). Simulating the joint evolution of quasars, galaxies and their large-scale distribution. *Nature* **435**, 629–636.

Srianand, R., Petitjean, P., & Ledoux, C. (2000). The cosmic microwave background radiation temperature at a redshift of 2.34. *Nature* **408**, 931–935.

Staggs, S., Dunkley, J., & Page, L. (2018). Recent discoveries from the cosmic microwave background: A review of recent progress. *Reports on Progress in Physics* **81**, 044901.

Stanford, P. K. (2006). *Exceeding Our Grasp: Science, History, and the Problem of Unconceived Alternatives.* Oxford: Oxford University Press.

Stebbins, A. (1997). The CMBR Spectrum, preprint astro-ph/9705178.

Steckline, V. S. (1983). Zermelo, Boltzmann, and the recurrence paradox. *American Journal of Physics* **51**, 894–897.

Steigman, G. (1978). A crucial test of the Dirac cosmologies. *Astrophysical Journal* **221**, 407–411.

Steigman, G., & Strittmatter, P. A. (1971). Neutrino limits on antimatter sources of energy in Seyfert galaxies. *Astronomy and Astrophysics* **11**, 279–285.

Strogatz, S. H. (2001). *Nonlinear Dynamics and Chaos: With Applications to Physics, Biology, Chemistry, and Engineering.* Boulder: Westview Press.

Sullivan, J. W. N. (1931). The physical nature of the universe. In W. Rose (ed.), *An Outline of Modern Knowledge.* London: Victor Golanz.

Sunyaev, R. A., & Zel'dovich, Ya. B. (1980). Microwave background radiation as a probe of the contemporary structure and history of the universe. *Annual Review of Astronomy and Astrophysics* **18**, 537–560.

Susskind, L. (2006). *The Cosmic Landscape: String Theory and the Illusion of Intelligent Design.* New York: Back Bay Books.

Thorne, K. S., & Will, C. M. (1971). Theoretical frameworks for testing relativistic gravity. I: Foundations. *The Astrophysical Journal* **163**, 595–610.

Tipler, F. J. (1982). Anthropic-principle arguments against steady-state cosmological theories. *Observatory* **102**, 36–39.

Tolman, R. C. (1934). *Relativity, Thermodynamics, and Cosmology.* Oxford: Clarendon Press.

Turner, M. S. (1992). The tilted universe. *General Relativity and Gravitation* **24**(1), 1–7.

Unzicker, A. (2009). A look at the abandoned contributions to cosmology of Dirac, Sciama, and Dicke. *Annalen der Physik* **521**, 57–70.

Vogelsberger, M., Genel, S., Springel, V., Torrey, P., Sijacki, D., Xu, D., Snyder, G., Nelson, D., & Hernquist, L. (2014). Introducing the Illustris Project: Simulating the coevolution of dark and visible matter in the Universe. *Monthly Notices of the Royal Astronomical Society* **444**(2), 1518–1547.

Wagoner, R. V., Fowler, W. A., & Hoyle, F. (1967). On the synthesis of elements at very high temperatures. *The Astrophysical Journal* **148**, 3–49.

Wainwright, C. L., Johnson, M. C., Peiris, H. V., Aguirre, A., Lehner, L., & Liebling, S. L. (2014). Simulating the universe (s): From cosmic bubble collisions to cosmological observables with numerical relativity. *Journal of Cosmology and Astroparticle Physics* **2014**(03), 030.

Wallace, A. R. (1903). *Man's Place in the Universe*. London: Chapman & Hall.

Weinberg, S. (1972). *Gravitation and Cosmology: Principles and Applications of the General Theory of Relativity*. New York: Wiley.

Weinberg, S. (2008). *Cosmology*. Oxford: Oxford University Press.

Wesson, P. S. (1975). The interrelationship between cosmic dust and the microwave background. *Astrophysics and Space Science* **36**, 363–382.

Wesson, P. S., & Seahra, S. S. (2001). Images of the Big Bang. *The Astrophysical Journal* **558**, L75–L78.

Weymann, R. (1966). The energy spectrum of radiation in the expanding universe. *Astrophysical Journal* **145**, 560–571.

Wheeler, J. A., & Feynman, R. P. (1945). Interaction with the absorber as the mechanism of radiation. *Review of Modern Physics* **17**, 157–161.

Wheeler, J. A., & Feynman, R. P. (1949). Classical electrodynamics in terms of direct interparticle action. *Review of Modern Physics* **21**, 425–433.

Whittaker, E. (1989). *A History of the Theories of Aether and Electricity: Vol. I: The Classical Theories; Vol. II: The Modern Theories, 1900–1926*. Mineola: Dover Publications.

Whitrow, G. J. (1962). Is the Physical Universe a Self-Contained System? *The Monist* 77–93.

Whitrow, G. J., & Bondi, H. (1954). Is physical cosmology a science? *The British Journal for the Philosophy of Science* **4**(16), 271–283.

Wickramasinghe, N. C., Edmunds, M. G., Chitre, S. M., Narlikar, J. V., & Ramadurai, S. (1975). A dust model for the cosmic microwave background. *Astrophysics and Space Science* **35**, L9–L13.

Will, C. M. (1971). Theoretical frameworks for testing relativistic gravity. II: Parametrized post-Newtonian hydrodynamics, and the Nordtvedt effect. *The Astrophysical Journal* **163**, 611–628.

Will, C. M., & Nordtvedt, K. Jr. (1972). Conservation laws and preferred frames in relativistic gravity. I: Preferred-frame theories and an extended PPN formalism. *The Astrophysical Journal* **177**, 757–774.

Willman, B., Blanton, M. R., West, A. A., Dalcanton, J. J., Hogg, D. W., Schneider, D. P., Wherry, N., Yanny, B., & Brinkmann, J. (2005). A new Milky Way companion: Unusual globular cluster or extreme dwarf satellite? *The Astronomical Journal* **129**(6), 2692–2700.

Winsberg, E. (2019). *Science in the Age of Computer Simulation*. Chicago: University of Chicago Press.

Woit, P. (2011). *Not Even Wrong: The Failure of String Theory and the Continuing Challenge to Unify the Laws of Physics*. New York: Random House.

Wolfe, A. M., & Burbidge, G. R. (1969). Discrete source models to explain the microwave background radiation. *Astrophysical Journal* **156**, 345–371.

Woody, D. P., & Richards, P. L. (1978). Spectrum of the cosmic background radiation. Preprint. Available online at https://escholarship.org/uc/item/5wt0g1v6 (last accessed June 20, 2023).

Woody, D. P., & Richards, P. L. (1979). Spectrum of the cosmic background radiation. *Physical Review Letters* **42**(14), 925–929.

Wright, E. L. (1982). Thermalization of starlight by elongated grains: Could the microwave background have been produced by stars? *Astrophysical Journal* **255**, 401–407.

Wright, E. L. (1987). Source counts in the chronometric cosmology. *Astrophysical Journal* **313**, 551–555.

Wright, E. L. (2003). The WMAP data and results. *New Astronomy Reviews* **47**, 877–881.

Wright, E. L., et al. (1994). Interpretation of the COBE FIRAS CMBR spectrum. *Astrophysical Journal* **420**, 450–456.

Yi-Fu, C., Easson, D., & Brandenberger, R. (2012). Towards a nonsingular bouncing cosmology. *Journal of Cosmology and Astroparticle Physics* **8**, 020.

Zel'dovich, Ya. B. (1963). Star production in an expanding universe. *Journal of Experimental and Theoretical Physics* **16**, 1395–1396.

Zel'dovich, Ya. B. (1964). The theory of the expanding universe as originated by A. A. Fridman. *Soviet Physics Uspekhi* **6**, 475–494.

Zel'dovich, Ya. B. (1972). A hypothesis, unifying the structure and the entropy of the universe. *Monthly Notices of the Royal Astronomical Society* **160**, 1–3.

Zollman, K. J. (2007). The communication structure of epistemic communities. *Philosophy of Science* **74**(5), 574–587.

Zwier, K. R. (2013). An epistemology of causal inference from experiment. *Philosophy of Science* **80**(5), 660–671.

Index

Printed in the United States
by Baker & Taylor Publisher Services